SOLUTIONS MANUAL

FOR

MODERN GENETIC ANALYSIS

William D. Fixsen

Diane K. Lavett

INTERACTIVE GENETICS WITH FREE CD-ROM

Lianna Johnson

John Merriam

W. H. Freeman and Company
New York

ISBN: 0-7167-9827-1 (EAN: 9780716798279)

Printed in the United States of America

Third printing

W. H. Freeman and Company
41 Madison Avenue
New York, NY 10010

www.whfreeman.com

CONTENTS

PREFACE

It is my hope that this manual will provide useful explanations, numerous approaches to solving problems, and, of course, correct answers. However, having taught genetics for many years to thousands of students, it is my belief that focusing too much on the latter misses the point of the exercise. In the end, learning to recognize what steps you actually take, what knowledge and logic you apply, is more important than the answer. Certainly getting the correct answer is satisfying and should be considered the reward for all your hard work, but understanding how you solved the problem will utimately be more useful. So when attempting to solve a problem, do not give up too easily and do not fool yourself by just looking at the answer and thinking "right, that would have been my answer too."

During my own education, I have had a number of teachers and mentors who have passed on their knowledge and love for genetics to whom I am indebted. Dr. Elizabeth Jones, Dr. David Botstein, Dr. Maurice Fox, and Dr. H. Robert Horvitz have all played key roles in my genetic training; I hope that I am able to convey some of their enthusiasm in my own work. I would also like to thank Dr. William Gelbart for recommending me for this project.

William D. Fixsen
*Department of Molecular
and Cellular Biology
Harvard University*

To my grandest genetic experiments,
Allie and Jessie

1 GENETICS AND THE ORGANISM

1. *Genetics* is the study of genes and genomes: their biochemical basis; how they function; how they are controlled; how they are organized; how they replicate; how they change; how they can be manipulated; and how they are transmitted from cell to cell and generation to generation.

   The ancient Egyptian racehorse breeders can be only loosely classified as geneticists because their interests were highly focused on producing fast horses, rather than on attempting to understand the mechanisms of heredity. Their understanding of the processes involved in producing fast horses was very incorrect and their methods were not analytic in the modern sense of the word. Nevertheless, they did produce very fast horses through a combination of observation, trial and error, and artificial selection.

2. DNA determines all the specific attributes of a species (shape, size, form, behavioral characteristics, biochemical processes, etc.) and sets the limits for possible variation that is environmentally induced.

3. Properties of DNA that are vital to its being the hereditary molecule are: its ability to replicate; its informational content; and its relative stability while still retaining the ability to change or mutate. Alien life forms might utilize RNA, just as some viruses do, as a hereditary molecule. However, of the types of molecules that can exist on earth, only the nucleic acids possess the necessary characteristics.

4. There are four possible nucleotide pairs at each position (A–T, T–A, G–C, or C–G). Therefore, the general formula for calculating the different possibilities is 4^n (where n = the number of pairs). In this case, there are 4^{10} or 1,048,576 possible DNA molecules!

5. There are many ways to simply draw DNA. The point of this exercise is to realize that in double-stranded DNA, the two strands are antiparallel, meaning that the sugars of each are oriented in opposite directions, and the bases associate by hydrogen bonds to hold the two strands together.

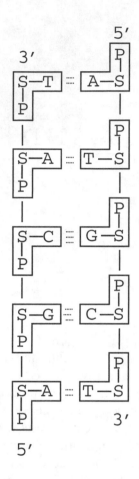

6. a. Human somatic cells are diploid (2*n*) and the haploid number is 23 (*n* = 23). Each chromosome is a double-stranded DNA molecule, so there are 46 molecules of DNA.

   b. It is generally stated that the haploid number represents the number of different types of DNA molecules, but that does not take into account the difference between the X and Y chromosomes in the different sexes. So females have 23 different types (called 1–22 and X) and males have 24 (1–22, X, and Y).

7. Keeping this strand antiparallel to the strand given, the sequence would be:

3´ — TAACCACGTAATGAAGTCCGAGA — 5´

8. Yes. There are no sequence restrictions in single-stranded DNA. The percentage of A must equal the percentage of T only in double-stranded DNA.

9. **a.** No.

    b. Yes, since A = T and G = C, the equation A + C = G + T can be rewritten as T + C = C + T by substituting the equal terms.

10. For the following, normal typeface represents previously polymerized nucleotides and *italic* typeface represents newly polymerized nucleotides.

    3´ — TTGGCACGTCGTAAT — 5´
    5´ — *AACCGTGCAGCATTA* — 3´

    3´ — *TTGGCACGTCGTAAT* — 5´
    5´ — AACCGTGCAGCATTA — 3´

11. Remember, the transcript will be antiparallel to the DNA template.

    3´ — UUGGCACGUCGUAAU — 5´

12. Sulfur. The sugar contains carbon, oxygen, and hydrogen; the phosphate group contains phosphorus and oxygen; and the base contains carbon, oxygen, hydrogen, and nitrogen.

13. The simplest definition is that a *gene* is a chromosomal region capable of making a functional transcript. However, this does not take into account the regulatory regions near the gene necessary for the proper expression of the gene nor the regions that help control transcription that can be quite distant. Also, many eukaryotes have large regions of noncoding sequences (introns) interspersed within the regions that encode the product (exons).

2 THE STRUCTURE OF GENES AND GENOMES

1. There are a number of ways of connecting these terms. One might be:

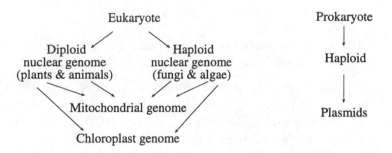

2. In each cell there are two copies of the nuclear genome (plants are diploid) and many copies of the mitochondrial and chloroplast genomes.

3. Diploid organisms have two copies of each of their different chromosomes. The term homologous is used to refer to the two members of each pair. Homologous chromosomes are alike in that they carry the same genes in the same relative positions. However, these genes may differ in informational content (representing different alleles).

4. A *chromosome* is a single DNA molecule containing many genes, often with associated protein and RNA.

A *chromomere* is a small, beadlike structure visible on a chromosome during prophase of mitosis and meiosis.

A *chromocenter* is the point at which polytene chromosomes appear to be attached together.

Chromatin is the substance of chromosomes; it includes the DNA, and associated proteins and RNA.

5. **a.** The following have mitochondria: a fish, moss, a palm tree, and bakers' yeast.

   b. The following have chloroplasts: a diatom and mistletoe.

6. The DNA double helix is held together by two types of bonds, covalent and hydrogen. Covalent bonds occur within each linear strand and strongly bond the bases, sugars, and phosphate groups (both within each component and between components). Hydrogen bonds occur between the two strands and involve a base from one strand with a base from the second in complementary pairing. These hydrogen bonds are individually weak but collectively quite strong.

7. If the DNA is double-stranded, A = T and G = C and A + T + C + G = 100%. If T = 15%, then C = [100 − 15(2)]/2 = 35%.

8. If the DNA is double-stranded, the G content should equal the C content. Since the total content cannot equal more than 100%, it is not possible to have 55% G. (However, it is possible if the DNA is single-stranded.)

9. A sequence would be most meaningful if it included both the order of bases and the polarity of the strand. Whether this sequence is written 5′ to 3′ or 3′ to 5′ would alter its information content.

10. **a.** Since diploid organisms usually have two copies of each of their chromosomes, an odd number of chromosomes likely indicates the organism is haploid.

    b. Although an odd number of chromosomes likely indicates the organism is haploid, an even number does not necessarily indicate the organism is diploid. Either this organism is haploid with 6 different chromosomes, or it is diploid with 3 pairs of chromosomes.

11. Since this DNA molecule of 1000 base pairs is 60% G+C, 600 base pairs would have 3 hydrogen bonds each (for a total of 1800) and 400 (40% A+T) would have 2 hydrogen bonds each (for a total of 800). Thus, there would be a total of 2600 hydrogen bonds holding the two strands of this DNA molecule together.

12. No. The average size of genes varies from one organism to another because the number and length of introns and the amount of repetitive DNA varies greatly.

13. The choices are many, but the following is one example. Nucleosomes, consisting of DNA wrapped around a core of histone proteins, associate to form a coiled structure called a solenoid, as chromosomes are more tightly packaged.

14. If the DNA is double-stranded, G = C = 24% and A = T = 26%.

15. There are many ways to indicate the polarity of DNA in a simple way. The point of this exercise is to realize that the polarity is based on how the sugar (deoxyribose) is oriented within the backbone of each strand. The 5´ carbon is attached to a phosphate while the 3´ carbon has an hydroxyl group to which new nucleotides may be added. In double-stranded DNA, the two strands are antiparallel, meaning that the sugars of each are oriented in opposite directions.

16. Human somatic cells are diploid (*2n*) and the haploid number is 23 (*n* = 23).

17. 46 DNA molecules (chromosomes)

18. 3´ – TAACCGAGA – 5´

19. a. Yes. Since A = T and G = C, the equation A + C = G + T can be rewritten as T + C = C + T by substituting the equal terms.

b. Yes. The equation A + G = C + T can be rewritten as T + C = C + T by substituting the equal terms.

c. Yes, the percentage of purines will equal the percentage of pyrimidines in double-stranded DNA. This is stating in words what is also shown in part b.

d. Ignoring the possibility of each 5´ end still having three phosphates attached, the DNA backbone is an alternating sugar-phosphate structure, and therefore the number of phosphates equals the number of deoxyribose sugars.

20. a. Sulfur is not found in DNA.

b. Neither nitrogen nor sulfur is found in the backbone of DNA.

21. The simplest definition is that a *gene* is a chromosomal region capable of making a functional transcript. However, this does not take into account the regulatory regions near the gene necessary for the proper expression of the gene nor the regions that help control transcription that can be quite distant. Also, many eukaryotes have large regions of noncoding sequences (introns) interspersed within the regions that encode a product (exons).

22. **a.** 1830 kb/1703 genes = 1074 base pairs

   b. 1074 – 1000 = 74 base pairs

   c. These DNA sequences will predominantly be regulatory regions.

   d. 0%. Bacteria do not have introns.

   e. The same as part b, 74 base pairs.

23. Many eukaryotes have intervening sequences called *introns*, which are transcribed but then spliced out (removed) prior to translation. In this case, the 25-kb primary transcript has all but 2.1 kb removed to generate the proper mRNA necessary for translation.

24. PFGE separates DNA molecules by size. When DNA is carefully isolated from *Neurospora* (which has 7 different chromosomes), 7 bands should be produced using this technique. Similarly, the pea has 7 different chromosomes and will produce 7 bands (homologous chromosomes will co-migrate as a single band). The housefly has 6 different chromosomes and should produce 6 bands.

25. mRNA size = gene size – (number of introns × average size of introns) – size of regulatory region

26. $(1.7\%) \times 3 \times 10^9 = 51 \times 10^6$ nucleotide differences

27. **a.** The purine bases are the larger, two-ring (planar, aromatic, heterocyclic) structures while the pyrimidines are the smaller one-ring structures. Thus, each purine/pyrimidine base pair is of equal width.

   b. Adenine and guanine are both purines; however, the chemical structure of guanine (with one keto and one amino group) allows for three hydrogen bonds when paired with cytosine, while adenine (with one amino group) forms two hydrogen bonds when paired with thymine.

   c. Cytosine and thymine are both pyrimidines; however, thymine has one methyl and two keto groups on its ring, while cytosine has one amino and one keto group on its ring.

28. The polarity of the sugar-phosphate backbone in DNA is defined by the orientation of the deoxyribose sugar. By convention, the carbons are numbered as primes to differentiate the atoms of the base from those of the sugar. It is useful to use 5´ and 3´ to define the orientation of DNA (or RNA) strands. In this way, the two strands of the double helix are said to be antiparallel (run in opposite orientations). Also, as you learn more about the enzymology of replication, transcription, and translation, the importance and concept of strand orientation can easily be conveyed.

29.

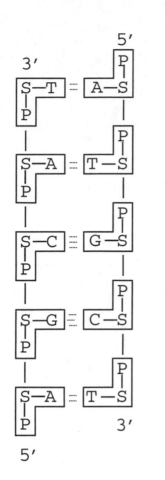 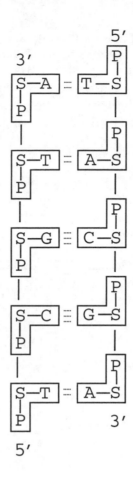

30. There are roughly 10 base pairs per turn.

31. 9 polytenes. Each homologous pair would be joined as one.

32. Herpes virus < *E. coli* < *Arabidopsis thaliana* < *Homo sapiens*

33. G bands are rich in A-T base pairs; R bands are rich in G-C base pairs. Staining with mithramycin would produce R bands.

34. Taking the average number of genes from chromosomes 21 and 22 and assuming this to be representative of all human chromosomes, the total number of genes in the human genome would be: (225 + 545)/2 = 385 genes per 1% of the genome or 38,500 for the entire genome. (This is in line with the predicted 30,000 to 40,000 genes suggested by the preliminary completion of the sequencing of the human genome.) An obvious source of error in this calculation is the simple assumption that genes are spread proportionally over the entire genome. Given that chromosome 22 contains more than twice as many genes as chromosome 21, it is likely that this assumption is too simple.

35. Since chromosome 22 is 33.46 mb and contains 14 mb of repetitive DNA elements, about 42% (14/33.46 = 0.418) of the chromosome is comprised of repetitive DNA.

36. The average amount of DNA per gene is 33.46 mb/545 genes = 61.4 kb. Assuming the gene has 5 introns, it would have 6 exons of average length 266 bp or approximately 1.6 kb of coding DNA (6 × 266 = 1596 bp) and each intron would be, on average, 3.52 kb long (19.2 kb – 1.6 kb = 17.6 kb of introns divided by 5 = 3.52 kb per intron). The average spacing between genes would be 61.4 kb (the average amount of DNA per gene) – 19.2 kb (the average gene size) = 42.4 kb.

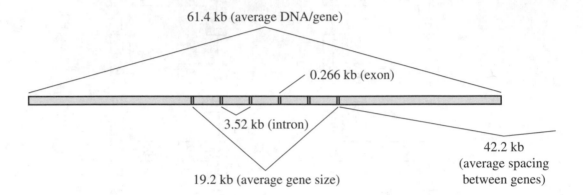

61.4 kb (average DNA/gene)

0.266 kb (exon)

3.52 kb (intron)

42.2 kb (average spacing between genes)

19.2 kb (average gene size)

3 GENE FUNCTION

1. **a.** By studying the genetic code table provided in the textbook, you will discover that there are eight cases in which knowing the first two nucleotides does not tell you the specific amino acid.

    b. If you knew the amino acid, you would not know the first two nucleotides in the cases of Arg, Ser, and Leu.

2.   3´ CGT   ACC   ACT   GCA 5´   DNA double helix (transcribed strand)

        5´ GCA   TGG   TGA   CGT 3´   DNA double helix

        5´ GCA   UGG   UGA   CGU 3´   mRNA transcribed

        3´ CGU   ACC   ACU   GCA 5´   Appropriate tRNA anticodon

        NH_3 - Ala - Trp - (stop) - COOH   Amino acids incorporated

3. **a., b.** 5´ UUG GGA AGC 3´

    c., d. Assuming the reading frame starts at the first base:
        NH_3 - Leu - Gly - Ser - COOH

    For the bottom strand, the mRNA is 5´ GCU UCC CAA 3´ and, assuming the reading frame starts at the first base, the corresponding amino acid chain is NH_3 - Ala - Ser - Gln - COOH.

4. **a.** *his⁻* (since it cannot make histidine).

    b. *y* (lowercase because it's recessive).

    c. *Pr* (uppercase because it's dominant).

d. *H* (uppercase because it's dominant). Although most null mutations are recessive, some are dominant because of haplo-insufficiency.

e. *y⁺/y; Pr/Pr⁺; H/H⁺*

5. Transcription (mRNA/DNA), translation (tRNA/mRNA), translation initiation (rRNA/mRNA), splicing (snRNP/mRNA)

6. There are many: polymerases, nucleases, transcription factors, ligases, restriction enzymes, ribosomal proteins, histones, topoisomerases, etc.

7. **a., b., c., d., e., f.**

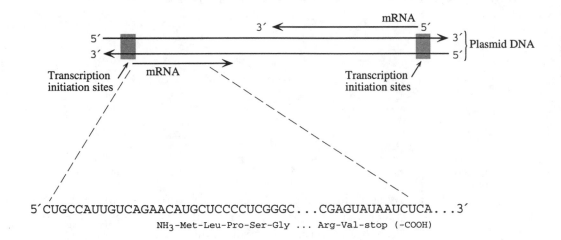

8. **a.** The main use is in detecting carrier parents and in diagnosing the disorder in the fetus.

   b. Because the values for normal individuals and carriers overlap for galactosemia, there is ambiguity if a person has 25 to 30 units. That person could be either a carrier or normal.

   c. These wild-type genes are phenotypically dominant but are incompletely dominant at the molecular level. A minimal level of enzyme activity apparently is enough to ensure normal function and phenotype.

9. **a.** If enzyme A was defective or missing (*m2/m2*), red pigment would still be made and the petals would be red.

   b. Purple, because it has a wild-type allele for each gene and you are told that the mutations are recessive.

   c. The mutant alleles do not produce functional enzyme. However, enough functional enzyme must be produced by the single wild-type allele of each gene to synthesize normal levels of pigment.

10. **a.** If enzyme B is missing, a white intermediate will accumulate and the petals will be white.

b. If enzyme D is missing, a blue intermediate will accumulate and the petals will be blue.

11. **a.** Null alleles are mutations that destroy the function of the protein. Missense, nonsense, deletion, and frameshift mutations within exons may all result in loss-of-function alleles. Although transcribed, the polypeptide may or may not be translated (or recognized by the experimenter) depending on the type of mutation.

b. There are sequences near the boundaries of and within introns that are necessary for correct splicing. If these are altered by mutation, correct splicing will be disrupted. Although transcribed, it is likely that translation will not occur.

c. The promoter is required for the proper initiation of transcription. Null alleles of the promoter will block transcription.

d. Yes; transcript yes; polypeptide no (see part b).

12. **a.** Recessive. The normal allele provides enough enzyme to be sufficient for normal function (the definition of haplo-sufficient).

b. There are many ways to mutate a gene to destroy enzyme function. One possible mutation might be a frameshift mutation within an exon of the gene. Assuming that a single base pair was deleted, the mutation would completely alter the translational product 3´ to the mutation.

c. Hormone replacement could be given to the patient.

d. If the hormone is required before birth, it can be supplied by the mother.

13. Former nucleotide triphosphates have just liberated inorganic diphospoate at the 3´ ends of the newly synthesized RNAs. RNA, like DNA, is polymerized 5´ to 3´, so the most recently added nucleotides in this figure will be the U found at the 3´ end of the transcript of gene 1 and and C found at the 3´ end of the transcript of gene 2.

14. DNA has often been called the "blueprint of life"—but how does it actually compare to a blueprint used for house construction? Both are abstract representations of instructions for building three-dimensional forms and both require interpretation for their information to be useful. But real blueprints are two-dimensional renderings of the various views of the final structure drawn to scale. There is one-to-one correlation between the lines on the drawing and the real form. The information in the DNA is encoded in a linear array—a one-dimensional set of instructions that only becomes three-dimensional as the encoded linear array of amino acids fold into their many forms. Included in the informational content of the DNA are also all the directions required for its "house" to maintain and repair itself, respond to change, and replicate. Let's see a real blueprint do that!

15. Other than the human ability to synthesize and wear synthetic materials, living systems are "proteins or something that has been made by a protein."

16. There are three codons for isoleucine: 5´ AUU 3´, 5´ AUC 3´, and 5´ AUA 3´. Possible anticodons are 3´ UAA 5´ (complementary), 3´ UAG 5´ (complementary), and 3´ UAI 5´ (wobble). 5´ UAU 3´, although complementary, would also basepair with 5´ AUG 3´ (methionine) due to wobble and therefore would not be an acceptable alternative.

17. **a.** The data cannot indicate whether one or both strands are used for transcription. You do not know how much of the DNA is transcribed or which regions of DNA are transcribed.

    b. If the RNA is double-stranded, the percentage of purines (A + G) would equal the percentage of pyrimidines (U + C) and the (A + G)/(U + C) ratio would be 1.0. This is clearly not the case for *E. coli*, which has a ratio of 0.80. The ratio for *B. subtilis* is 1.02. This is consistent with the RNA being double-stranded but does not rule out single-stranded if there are an equal number of purines and pyrimidines in the strand.

18. Growth will be supported by a particular compound if it is later in the pathway than the enzymatic step blocked in the mutant. Restated, the more mutants a compound supports, the later in the pathway it must be. In this example, compound G supports growth of all mutants and can be considered the end product of the pathway. Alternatively, compound E does not support the growth of any mutant and can be considered the starting substrate for the pathway. The data indicate the following:

    a., b.

    $$E \xrightarrow{\quad} A \xrightarrow{\quad} C \xrightarrow{\quad} B \xrightarrow{\quad} D \xrightarrow{\quad} G$$
    $$\quad 5 \qquad 4 \qquad 2 \qquad 1 \qquad 3$$

    vertical lines indicate step where each mutant is blocked

19. There are a number of explanations that might explain the positional clustering of null mutants in this *Neurospora* gene. However, based on the material covered in this chapter, it is possible that the gene codes for an enzyme that uses its central amino acids (encoded by the central region of the gene) to form its active site. Mutations that alter the active site amino acids are most likely to destroy enzymatic function and would be classified as null (loss-of-activity) mutations.

20. Protein function can be destroyed by a mutation that causes the substitution of a single amino acid even though the protein has the same immunological properties. For example, enzymes require very specific amino acids in exact positions within their active site. A substitution of one of these key amino acids might have no effect on overall size and shape of the protein while completely destroying the enzymatic activity.

21. **a.** Each wild-type allele produces 10 units of enzyme E per cell. A null mutation in the gene coding for this enzyme produces 0 units. More than 10

units of enzyme are required by the cells for wild-type levels of function and phenotype.

E^+/E^+	20 units	Wildtype
E^+/E	10 units	Mutant
E/E	0 units	Mutant

b. Each wild-type allele produces 15 units of enzyme F per cell. A null mutation in the gene coding for this enzyme produces 0 units. 15 units of enzyme are sufficient for wild-type phenotype.

f^+/f^+	30 units	Wild-ype
f^+/f	15 units	Wildtype
f/f	0 units	Mutant

22. a. The allele *sn* will show dominance over *sf* because there will be only 40 units of square factor in the heterozygote.

b. No. Here the functional allele is recessive.

c. The allele *sf* may become dominant over time by several mechanisms: it could mutate in a way that it produces 50 units, other genes might mutate to lower the cells needed to produce square phenotype to less than 50 units, or other genes might mutate to increase the production or activity of *sf.*

23. a. The mothers had an excess of phenylalanine in their blood, and that excess was passed through the placenta into the fetal circulatory system, where it caused brain damage prior to birth.

b. The diet had no effect because the neurological damage has already happened *in utero* prior to birth.

c. A fetus with two mutant copies of the allele that causes PKU makes no functional enzyme. However, the mother of such a child is heterozygous and makes enough enzyme to block any brain damage; the excess phenylalanine in the fetal circulatory system enters the maternal circulatory system and is processed by the maternal gene product. After birth, which is when PKU damage occurs in a PKU child, dietary restrictions block a buildup of phenylalanine in the circulatory system until brain development is completed.

The fetus of a PKU mother is exposed to the very high level of phenylalanine in its circulatory system during the time of major brain development. Therefore, brain damage occurs before birth, and no dietary restrictions after birth can repair that damage.

d. The solution to the brain damage seen in the babies of PKU mothers is to return the mother to a restricted diet during pregnancy in order to block high levels of exposure to her child.

e. PKU is characterized as a rare recessive disorder. A child with PKU has two parents that carry a mutant allele for the metabolism of phenylalanine. When two individuals who are heterozygous for PKU have a child, the risk that the child will have PKU is 25%. A PKU child is unable to make a functional enzyme that converts phenylalanine to tyrosine. As a result, an excess level of phenylalanine is found in the blood, and the excess is detected as an increase in phenylpyruvic acid in both the blood and the urine. The excess phenylpyruvic acid blocks normal development of the brain, resulting in retardation.

24. a., b. The goal of this type of problem is to align the two sequences. You are told that there is a single nucleotide addition and single nucleotide deletion, so look for single base differences that effect this alignment. These should be located where the protein sequence changes (i.e., between Lys-Ser and Asn-Ala). Remember also that the genetic code is redundant. (N = any base)

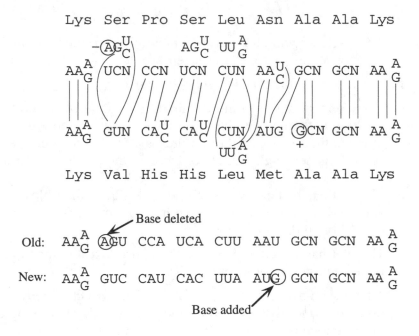

25. Mutant 1: null mutation; no product is made.

Mutant 2: leaky mutation; not enough product is made for normal function.

Mutant 3: nonsense mutation/frameshift; a truncated protein product is made.

Mutant 4: substitution mutation; protein is the same size but a single amino acid change destroys its function.

4 THE TRANSMISSION OF DNA AT CELL DIVISION

1. A **primer** is a short segment of RNA that is synthesized by primase, using DNA as a template during DNA replication. Once the primer is synthesized, DNA polymerase then adds DNA to the 3´ end of the RNA. Primers are required because the major DNA polymerase involved with DNA replication is unable to initiate DNA synthesis and, rather, requires a 3´ end. The RNA is subsequently removed and replaced with DNA so that no gaps exist in the final product.

2. The choices are many, but the following is one example. During replication, because of the antiparallel nature of DNA, the synthesis of new DNA is continuous for the leading strand and discontinuous for the lagging strand.

3. **a.** Prior to the S phase, each chromosome has two telomeres so in the case of $2n = 14$, there are 14 chromosomes and 28 telomeres.

   b. After S, each chromosome consists of two chromatids, each with two telomeres for a total of four telomeres per chromosome. So for 14 chromosomes, there would $14 \times 4 = 56$ telomeres.

   c. At prophase, the chromosomes still consist of two chromatids each, so there would $14 \times 4 = 56$ telomeres.

   d. At telophase, there would be 28 telomeres in each of the soon-to-be daughter cells.

4. Yes, both strands serve as templates along their entire length. For each, however, telomerase adds additional sequences to the 3´ end using its own RNA template, and these added sequences serve as templates for the newly synthesized lagging strands.

5. Both transcription and replication proceed 5´ to 3´ and antiparallel to the template and use triphosphates as building blocks. However, replication requires a primer, uses deoxyribonucleotides, and proceeds bidirectionally using both strands as templates, while transcription does not require a primer, uses ribonucleotides, and proceeds in one direction using just one strand as a template.

6. Six. The first replication start would have two replication forks proceeding to completion, and the now replicated origins would each start replication again and each would have two more replication forks for a total of six.

7. Only the DNA molecule that used the poly-T strand as a template would be radioactive. The other daughter molecule would not be radioactive since it would not have required any dATP for its replication.

Since each strand of the second molecule contains T, both daughter molecules would require dATP for replication, so each would be radioactive.

8. Graph of DNA content during mitosis and then meiosis:

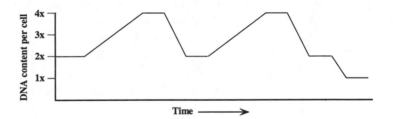

9. Because the DNA polymerase is capable of adding new nucleotides only at the 3´ end of a DNA strand, and because the two strands are antiparallel, at least two molecules of DNA polymerase must be involved in the replication of any specific region of DNA. When a region becomes single-stranded, the two strands have an opposite orientation. Imagine a single-stranded region that runs from right to left. The 5´ end is at the right, with the 3´ end pointing to the left; synthesis can initiate and continue uninterrupted toward the right end of this strand. Remember: new nucleotides are added in a 5´ → 3´ direction, so the template must be copied from its 3´ end. The other strand has a 5´ end at the left with the 3´ end pointing right. Thus, the two strands are oriented in opposite directions (antiparallel), and synthesis (which is 5´ → 3´) must proceed in opposite directions. For the leading strand (say, the top strand) replication is to the right, following the replication fork. It is continuous and may be thought of as moving "downstream." Replication on the bottom strand cannot move in the direction of the fork (to the right), since, for this strand, that would mean adding nucleotides to its 5´ end. Therefore, this strand must replicate discontinuously: as the fork creates a new single-stranded stretch of DNA, this is replicated *to the left* (away from the direction of fork movement). For this lagging strand, the replication fork is always opening new single-stranded DNA for replication *upstream* of the previously replicated stretch, and a new fragment of DNA is replicated back to the previously created fragment. Thus, one (Okazaki) fragment follows the other in the direction of the replication fork, but each fragment is created in the opposite direction.

10. As cells divide mitotically, each chromosome consists of identical sister chromatids that are separated to form genetically identical daughter cells. Although the second division of meiosis appears to be a similar process, the "sister" chromatids are likely to be different. Recombination during earlier meiotic stages has swapped regions of DNA between sister and non-sister chromosomes, such that the two daughter cells of this division are typically not genetically identical.

11.

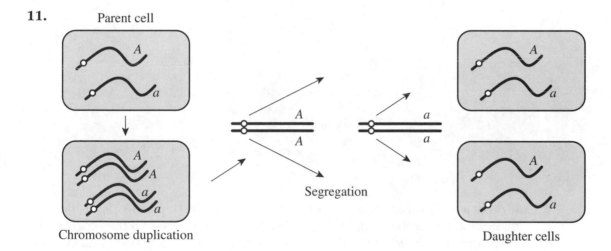

Parent cell

Chromosome duplication

Segregation

Daughter cells

12.

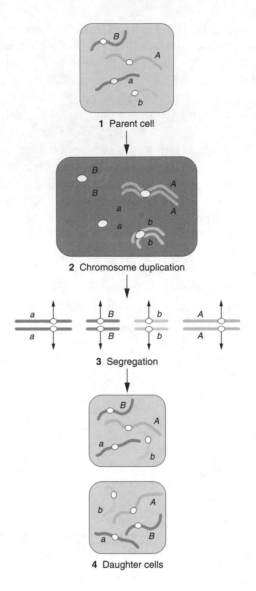

1 Parent cell

2 Chromosome duplication

3 Segregation

4 Daughter cells

13. If replication proceeded in a "conservative" manner, one DNA molecule after replication would be completely "old" or contain only ^{15}N, and the other would be completely "new" and contain only ^{14}N. Instead of seeing one band of intermediate density after centrifugation, Meselson and Stahl would have observed two bands.

14. The key function of mitosis is to generate two daughter cells genetically identical to the original parent cell. It could be argued that this "sameness" is conservative. The key functions of meiosis are to halve the DNA content and to reshuffle the genetic content of the organism to generate genetic diversity among the progeny. It could be argued that this "purposeful diversity" is liberal.

15. a. The bottom strand.

b.

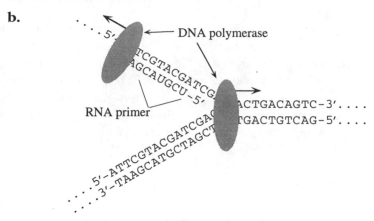

c.

```
....5'-ATTCGTACGATCGACTGACTGACAGTC-3'....
....3'-TAAGCATGCTAGCTGACTGACTGTCAG-5'....

....5'-ATTCGTACGATCGACTGACTGACAGTC-3'....
....3'-TAAGCATGCTAGCTGACTGACTGTCAG-5'....
```

d. Yes, but the other replication fork would be moving in the opposite direction, and the top strand, as drawn, would now be the leading strand, and the bottom strand would now be the lagging strand.

16. The nucleus contains the genome and separates it from the cytoplasm. However, during cell division, the nuclear envelope dissociates (breaks down) and it is the job of the microtubule-based spindle to actually separate the chromosomes (divide the genetic material) around which nuclei re-form during telophase. In this sense, it can be viewed as a passive structure that is divided by the cell's cytoskeleton.

17.

	Mitosis	Meiosis
fern	sporophyte gametophyte	sporophyte (sporangium)
moss	sporophyte gametophyte	sporophyte (atheridium and archegonium)
onion	sporophyte gametophyte	sporophyte (anther and ovule)
pine tree	sporophyte gametophyte	sporophyte (pine cone)
mushroom	sporophyte gametophyte	sporophyte (ascus or basidium)
frog	somatic cells	gonads
snail	somatic cells	gonads

18. Chromosome pairing is unique to prophase I of meiosis.

19. Recombination is another event that takes place in meiosis and not mitosis.

20. If the polymerase complex is moving from right to left and using the bottom strand as template, the new DNA will be replicated 5′ to 3′, so this represents the leading strand. As the top strand is used as the template, the polymerization of the new strand is in the opposite direction of the movement of the replication fork and would be the lagging strand. Thus the complement of the top strand would be the Okazaki fragment:

3′…GGAATTCTGATTGATGAATGACCCTAG…5′

21. Nucleotide triphphosphates are the building blocks for DNA. Ther energy released when inorganic diphosphate is liberated helps drive the polymerixation reaction. The remaining single phosphate still attached to each nucleotide makes up part of the backbone structure of the molecule. Only free nucleotides being added to the growing DNA chain will be in the triphosphate form and in this figure, only two of these are shown.

22. Without functional telomerase, the telomeres would shorten at each replication cycle leading to eventual loss of essential coding information and death. In fact, there are some current observations that decline or loss of telomerase activity plays a role in the mechanism of aging in humans.

23. The following drawing uses a solid line for the starting, nonradioactive DNA and a dotted line for the newly synthesized, radioactive DNA. The arrowheads are used to indicate the 3′ direction of the DNA strand and the oval represents the chromosome's centromere. For simplicity, a single chromosome is shown even though these cells are diploid. In (d), half the daughters will have DNA radioactive in both strands and half will have DNA that is half radioactive and half nonradioactive.

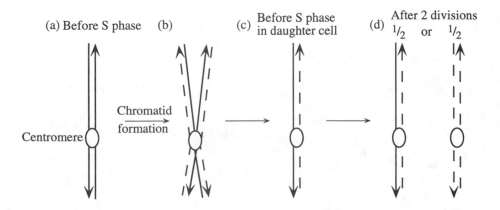

24. For an organism in which $2n = 6$, there are three chromosomes from each parent. To obtain a pollen grain containing all three chromosomes from a single parent requires the independent assorting of, in this case, the maternal centromeres during meiosis. The chance of this happening is $(1/2)^n$ or $1/8$. What would be the likelihood that you inherited a complete set of your paternal grandfather's centromeres?

25. Assuming that all cells are dividing randomly, then the proportion of cells in mitosis will represent that proportion of the cell cycle devoted to mitosis. If cells are not dividing randomly, or if all cells are not actively growing and dividing, then this method would not be informative.

26. It is likely that the failure of the spindle to attach to the kinetochore of just one chromatid would result in mitotic nondisjunction or chromosome loss.

27. Since the DNA levels vary four-fold, the range covers cells that are haploid (gametes) to cells that are diploid and dividing (after DNA has replicated but prior to cell division). The following cells would fit the DNA measurements:

33 units	haploid cells
66 units	diploid cells in G_1 or haploid cells after Meiosis I
132 units	diploid cells after S but prior to cell division

5

THE INHERITANCE
OF SINGLE-GENE DIFFERENCES

1. **a.** All the progeny will be heterozygous (*A/a*) and all will have the dominant A phenotype.

   b. If *A/a* is crossed to *a/a*, 50% of the progeny will be *A/a* and have the dominant A phenotype, and 50% will be *a/a* and have the recessive a phenotype.

   c. If *A/a* is crossed to *A/A*, all the progeny will have the dominant A phenotype, but 50% will be homozygous *A/A* and 50% will be heterozygous *A/a*.

   d. If *A/a* is crossed to *A/a*, 25% of the progeny will be homozygous *A/A* and phenotypically A, 50% will heterozygous *A/a* and phenotypically A, and 25% will be homozygous *a/a* and phenotypically a. In other words, a 1:2:1 genotypic ratio and a 3:1 phenotypic ratio.

2. **a.** Since all the progeny are orange, the orange phenotype is dominant to the red phenotype. Let *r* = allele for red and *R* = allele for orange.

The cross is:	Parents	$R/R \times r/r$	
	Progeny	all *R/r*	(orange)

   b. The cross is:

	Parents	$R/r \times R/r$	
	Progeny	25% *R/R*	(orange)
		50% *R/r*	(orange)
		25% *r/r*	(red)

3. Since all crosses with wild-type plants result in a 1:1 ratio of double to single, the interpretation is that these crosses are between heterozygous dominant plants and homozygous recessive ones. If the rare variant is a result of a new dominant mutation, then using *D* to represent the dominant double flowers allele and *d* to represent the recessive (and wild-type) single flowers allele, the crosses are:

Parents	$D/d \times d/d$	
Progeny	50% *D/d*	(double)
	50% *d/d*	(single)

4. First cousins share grandparents. If either grandparent was heterozygous for the recessive galactosemia allele (noted by shading in half of the symbol), the cousins could each inherit the allele from that grandparent.

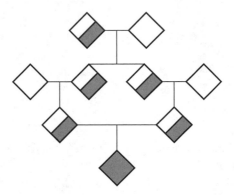

5. Do a testcross (cross to *a/a*). If the fly was *A/A*, all the progeny will be phenotypically A; if the fly was *A/a*, half the progeny will be A and half will be a.

6. Black (*B*) is dominant to white (*b*).

    Parents: $B/b \times B/b$

    Progeny: 3 black:1 white (1 *B/B*:2 *B/b*:1 *b/b*)

7. Charlie, his mate, or both, obviously were not pure-breeding, because his F_2 progeny were of two phenotypes. Let *A* = black and white, and *a* = red and white. If both parents were heterozygous, then red and white would have been expected in the F_1 generation. Red and white were not observed in the F_1 generation, so only one of the parents was heterozygous. The cross is:

P	$A/a \times A/A$
F_1	1 *A/a*:1 *A/A*

    Two F_1 heterozygotes (*A/a*) when crossed would give 1 *A/A* (black and white):2 *A/a* (black and white):1 *a/a* (red and white). If the red and white F_2 progeny were from more than one mate of Charlie's, then the farmer acted correctly. However, if the F_2 progeny came only from one mate, the farmer may have acted too quickly.

8.   *Unpacking the Problem*

a.

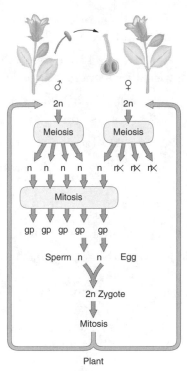

b.   *Dimorphic* means two forms. In this example, the plant has either blotched or unblotched leaves.

c.   The term *polymorphism* is more inclusive as it can also describe a situation where there are several (or many) phenotypic forms of a certain trait. Dimorphic is used when there are only two forms. For example, sexual phenotypes are dimorphic.

d.   A population is a group of organisms of the same species that are capable of interbreeding with one another.

e.   Your choice

f.   Not that were mentioned. Since the hypothesis being tested is that the phenotypic differences are the result of genotypic ones, it would be expected that any single plant would consist of cells of one genotype and therefore one phenotype.

g.   Leaf phenotype

h.   Two

i.   Blotched and unblotched

j.   No. The variation is likely the result of differences in phenotypic expression and would not affect the correct deduction of genotype.

k. Approximately the same

l. Since the plant had not flowered, pollination could be controlled in the lab and self-pollination could be insured.

m. Genotypes are often deduced by careful counting of progeny and statistical analysis of expectations and predicted results.

n. There are representative leaves of 120 progeny plants. Of these, 28 are from plants with unblotched leaves and 92 are from plants with blotched leaves.

o. Yes. There are 92:28 or 3.29:1 blotched:unblotched plants among the progeny.

p. Ratios are used to deduce genotypes and predict outcomes of certain genetic crosses.

q. 3:1; 1:1; 1:2:1

r. Yes. The progeny are approximately in a 3:1 ratio.

s. The phenotypic ratio is easily determined by observation. Genotypic ratios might be obscured by dominance. If one allele is completely dominant to another, heterozygotes and homozygotes for the dominant allele will have the same phenotype. In this example, there are only two phenotypes, but three genotypes. From the chapter, the cross $m^+/m \times m^+/m$ resulted in a 3:1 phenotypic ratio of normal:miniature winged fruit flies but a 1:2:1 genotypic ratio.

t. No. There may be many different alleles that result in the same phenotype. Also, in cases of partial or incomplete dominance, there may be more phenotypes than alleles.

u. Dominance describes the situation where the phenotype of the heterozygote is indistinguishable from that of one of the homozygotes.

v. "Allowed to self" means that the plant was mated to itself, that it provided both the sperm (pollen) and the eggs (ova). Many flowering plants produce both types of gametes and in some, these gametes are capable of self-fertilization. (Mendel made significant use of this in his studies with the garden pea.) In this example, a brush might be used to both collect and distribute pollen among the flowers of this single plant.

w. Progeny of different phenotype than the selfed parent indicate that the parent was heterozygous, that the parent's phenotype is dominant, and that the "new" progeny phenotype is recessive.

x. No. Since the population described contains both phenotypes, it is not correct to select one as wild type.

y. Typically, allele symbols are chosen that represent the non wild-type phenotype (sometimes called the mutant phenotype), and then upper and lowercase is used to designate dominance.

Solving the Problem

The plants are approximately 3 blotched:1 unblotched. This suggests that blotched is dominant to unblotched and that the original plant that was selfed was a heterozygote.

a. Let A = blotched, a = unblotched.

P A/a (blotched) × A/a (blotched)

F_1 1 A/A:2 A/a:1 a/a

3 $A/-$ (blotched):1 a/a (unblotched)

b. All unblotched plants should be pure-breeding in a testcross with an unblotched plant (a/a), and one-third of the blotched plants should be pure-breeding.

9. The results suggest that winged ($A/-$) is dominant to wingless (a/a) (cross 2 gives a 3:1 ratio). If that is correct, the crosses become

Pollination	Genotypes	Number of progeny plants	
		Winged	Wingless
Winged (selfed)	$A/A \times A/A$	91	1*
Winged (selfed)	$A/a \times A/a$	90	30
Wingless (selfed)	$a/a \times a/a$	4*	80
Winged × wingless	$A/A \times a/a$	161	0
Winged × wingless	$A/a \times a/a$	29	31
Winged × wingless	$A/A \times a/a$	46	0
Winged × winged	$A/A \times A/-$	44	0
Winged × winged	$A/A \times A/-$	24	0

The five unusual plants are most likely caused by either human error in classification or contamination. Alternatively, they could result from environmental effects on development. For example, too little water may have prevented the seed pods from becoming winged even though they are genetically winged.

10. **a.** The disorder appears to be dominant because all affected individuals have an affected parent. If the trait was recessive, then I-1, II-2, III-1, and III-8 would all have to be carriers (heterozygous for the rare allele).

b. Assuming dominance, the genotypes are

I: d/d, D/d

II: D/d, d/d, D/d, d/d

III: d/d, D/d, d/d, D/d, d/d, d/d, D/d, d/d

IV: D/d, d/d, D/d, d/d, d/d, d/d, d/d, D/d, d/d

c. The mating is $D/d \times d/d$. The probability of an affected child (D/d) equals $1/2$, and the probability of an unaffected child (d/d) equals $1/2$. Therefore, the chance of having four unaffected children (since each is an independent event) is: $1/2 \times 1/2 \times 1/2 \times 1/2 = 1/16$.

11. **a.** Autosomal recessive: affected individuals inherited the trait from unaffected parents and a daughter inherited the trait from an unaffected father.

b. Both parents must be heterozygous to have a $1/4$ chance of having an affected child. Parent 2 is heterozygous because her father is homozygous for the recessive allele, and Parent 1 has a $1/2$ chance of being heterozygous, because his father is heterozygous because 1's paternal grandmother was affected. Overall, $1 \times \frac{1}{2} \times \frac{1}{4} = \frac{1}{8}$.

12. **a.** *leu⁺/leu*

   b. The alleles will segregate during meiosis and the progeny will be: 1 *leu⁺*:1 *leu*.

   c. Certain amino acids are essential to protein structure and function; a change of even one of these could totally destroy an enzyme's activity. There are many ways to change a DNA sequence that encodes an enzyme, that will result in an altered amino acid sequence. For the following, a small DNA sequence has arbitrarily been chosen to show three classes of mutation that could cause loss of enzymatic activity.

      The first, a missense mutation, is the result of a single base pair change and alters a single amino acid. The second, a deletion of three base pairs, causes the loss of one amino acid. The third, a frameshift mutation caused by the addition of one base pair, alters the amino sequence from the site of the addition until the end of translation is reached. Other mutations, not shown, would also lead to null alleles: mutations in the promoter for this gene preventing transcription, mutations that alter or prevent proper splicing, insertions of DNA into the coding region of the gene, etc.

Wild-type allele

```
....5'-ATTCGTACGATCGAC-3'....
....3'-TAAGCATGCTAGCTG-5'....        DNA
           /              \
....5'-AUU CGU ACG AUC GAC-3'....     mRNA

....NH₂-Ile Arg Thr Ile Asp-COOH.... Protein
```

Missense allele

```
....5'-ATTC A TACGATCGAC-3'....
....3'-TAAG T ATGCTAGCTG-5'....       DNA
           /              \
....5'-AUU CAU ACG AUC GAC-3'....     mRNA

....NH₂-Ile His Thr Ile Asp-COOH.... Protein
```

Small deletion allele (3 pase pairs)

```
            CGT
            GCA
            ▽
....5'-ATTACGATCGAC-3'....
....3'-TAATGCTAGCTG-5'....            DNA
        /            \
....5'-AUU ACG AUC GAC-3'....         mRNA

....NH₂-Ile Thr Ile Asp-COOH....      Protein
```

Frameshift allele (+1)

```
....5'-ATTCGTA A CGATCGAC-3'....
....3'-TAAGCAT T GCTAGCTG-5'....      DNA
           /               \
....5'-AUU CGU AAC GAU CGA C-3'....   mRNA

....NH₂-Ile Arg Asn Asp Arg-COOH.... Protein
```

13. P s^+/s^+ × s/Y

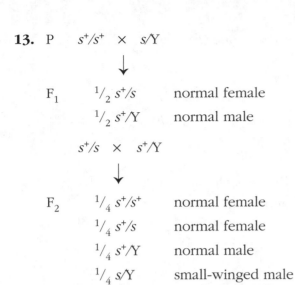

F$_1$ $^1/_2$ s^+/s normal female

        $^1/_2$ s^+/Y normal male

s^+/s × s^+/Y

F$_2$ $^1/_4$ s^+/s^+ normal female

        $^1/_4$ s^+/s normal female

        $^1/_4$ s^+/Y normal male

        $^1/_4$ s/Y small-winged male

In the cross: P s^+/s × s/Y

Progeny $^1/_4$ s^+/s normal female

        $^1/_4$ s/s small-winged female

        $^1/_4$ s^+/Y normal male

        $^1/_4$ s/Y small-winged male

14. Horizontal lines (H) is dominant to vertical lines (h):

        h/h × H/H

All H/h

Self $H/h \times H/h$

        $^3/_4$ ($^1/_4$ H/H + $^1/_2$ H/h)

        $^1/_4$ h/h

15. Black (B) is dominant to white (b):

Parents B/B × B/b

Progeny B/B × B/b × B/b × B/B

        $^1/_2$ B/b $^1/_2$ B/b $^1/_2$ B/B

        $^1/_2$ B/B $^1/_4$ B/B $^1/_2$ B/b

                $^1/_4$ b/b

16. Vertical line (h^+) is dominant to horizontal line (h):

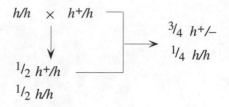

$$h/h \times h^+/h$$

$$\downarrow$$

$$\frac{1}{2}\ h^+/h$$
$$\frac{1}{2}\ h/h$$

$$\frac{3}{4}\ h^+/-$$
$$\frac{1}{4}\ h/h$$

17. Purple (B) is dominant to blue (b) and the trait is X-linked.

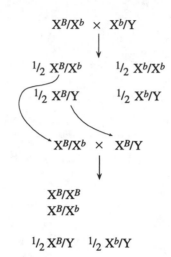

$$X^B/X^b \times X^b/Y$$

$$\downarrow$$

$$\frac{1}{2}\ X^B/X^b \qquad \frac{1}{2}\ X^b/X^b$$
$$\frac{1}{2}\ X^B/Y \qquad \frac{1}{2}\ X^b/Y$$

$$X^B/X^b \times X^B/Y$$

$$\downarrow$$

$$X^B/X^B$$
$$X^B/X^b$$

$$\frac{1}{2}\ X^B/Y \quad \frac{1}{2}\ X^b/Y$$

18. Star (s) is recessive to no star (s^+) and is X-linked.

$$X^s/X^s \quad \times \quad X^{s+}/Y$$

$$\downarrow$$

$$X^{s+}/X^s \quad \times \quad X^s/Y$$

$$\downarrow$$

$$\frac{1}{2}\ X^{s+}/X^s \qquad \frac{1}{2}\ X^s/X^s$$
$$\frac{1}{2}\ X^{s+}/Y \qquad \frac{1}{2}\ X^s/Y$$

19. You should draw pedigrees for this question.

a.

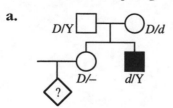

The "maternal grandmother" had to be a carrier, *D/d*. The probability that the woman inherited the *d* allele from her is $\frac{1}{2}$. The probability that she passes it to her child is $\frac{1}{2}$. The probability that the child is male is $\frac{1}{2}$. The total probability of the woman having an affected child is $\frac{1}{2} \times \frac{1}{2} \times \frac{1}{2} = \frac{1}{8}$.

b. The same pedigree as part **a** applies. The "maternal grandmother" had to be a carrier, *D/d*. The probability that your mother received the allele is $\frac{1}{2}$. The probability that your mother passed it to you is $\frac{1}{2}$. The total probability is $\frac{1}{2} \times \frac{1}{2} = \frac{1}{4}$.

c.

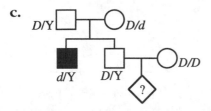

Because your father does not have the disease, you cannot inherit the allele from him. Therefore, the probability of inheriting an allele will be based on the chance that your mother is heterozygous. Since she is "unrelated" to the pedigree, assume that this is zero.

20. a. The pedigree indicates that the disease is caused by an X-linked recessive allele. You are told that the trait is rare, and in this pedigree only males are affected. Since both affected males have unaffected parents, the allele must be recessive. If autosomal, both parents of both affected males would have to be heterozygous, but since it is rare, it is more likely that the grandmother was heterozygous for an X-linked recessive allele that was inherited by one son and the second daughter, who subsequently passed the allele to one of her sons.

b. Generation I: X^+/Y, X^+/X^m

Generation II: X^+/X^+, X^m/Y, X^+/Y, $X^+/-$, X^+/X^m, X^+/Y

Generation III: X^+/X^+, X^+/Y, X^+/X^m, X^+/X^m, X^+/Y, X^+/X^+, X^m/Y, X^+/Y, $X^+/-$

c. The first couple has little chance of having an affected child because the male is normal and the chance of the unrelated female's being heterozygous is small. The second couple has a 50% chance of having affected sons and no chance of having affected daughters. The third couple has little chance of having an affected child (again, because the chance of the unrelated female's being heterozygous is rare), but all daughters will be heterozygous.

21. a. From cross 6, Bent (*B*) is dominant to normal (*b*). Both parents are "bent," yet some progeny are "normal."

b. From cross 1, it is X-linked. The trait is inherited in a sex-specific manner—all sons have the mother's phenotype.

c. In the following table, the Y chromosome is stated; the X is implied.

	Parents		Progeny	
Cross	Female	Male	Female	Male
1	b/b	B/Y	B/b	b/Y
2	B/b	b/Y	B/b, b/b	B/Y, b/Y
3	B/B	b/Y	B/b	B/Y
4	b/b	b/Y	b/b	b/Y
5	B/B	B/Y	B/B	B/Y
6	B/b	B/Y	B/B, B/b	B/Y, b/Y

22. a., b. The "stopper" or "continuous" phenotype of the maternal parent is inherited by all its offspring. This is typical of maternal inheritance (or organelle-based inheritance) and would be expected if this trait mapped to a mitochondrial gene.

Nuclear genes that differ between the two parental strains should segregate in the normal Mendelian manner and produce 1:1 ratios in the progeny. (Remember, *Neurospora* is haploid and the progeny of these crosses are actually the haploid products of meiosis.) This is observed for the orange/yellow trait, suggesting that this trait maps to a nuclear gene.

23. a. Galactosemia pedigree

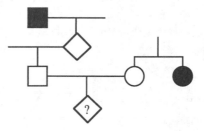

b. Both parents must be heterozygous for this child to have a $\frac{1}{4}$ chance of inheriting the disease. Since the mother's sister is affected with galactosemia, their parents must have both been heterozygous. Since the mother does not have the trait, there is a $\frac{2}{3}$ chance that she is a carrier (heterozygous). One of the father's parents must be a carrier since his grandfather had the recessive trait. Thus, the father had a $\frac{1}{2}$ chance of inheriting the allele from that parent. Since these are all independent events, the child's risk is

$$\frac{1}{4} \times \frac{2}{3} \times \frac{1}{2} = \frac{1}{12}$$

c. If the child has galactosemia, both parents must be carriers and thus those probabilities become 100%. Now all future children have a $\frac{1}{4}$ chance of inheriting the disease.

24. a. Favism pedigree

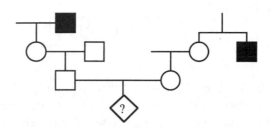

b. You are told that favism is caused by an X-linked recessive allele. The father does not have favism, so he must have the wild-type allele. Thus, for their child to inherit this trait, the mother must be a carrier (heterozygous) and the child must be male. The mother's uncle has favism, so we can assume that her maternal grandmother was heterozygous for the allele. In that case, there is a $\frac{1}{2}$ chance that her mother inherited the allele and another $\frac{1}{2}$ chance that she passed it to her. Since these are independent, there is a $\frac{1}{4}$ chance that the mother is a carrier. Finally, if she is a carrier, there is a $\frac{1}{2}$ chance she will pass the allele to her child and $\frac{1}{2}$ chance that the child is male:

$$\frac{1}{4} \times \frac{1}{2} \times \frac{1}{2} = \frac{1}{16}$$

c. Once it is determined that the mother is a carrier, the chance of passing the allele to the next child is still $\frac{1}{2}$, as is the chance of that child being male, so $\frac{1}{2} \times \frac{1}{2} = \frac{1}{4}$.

25. a. Recessive. Affected individuals inherited the trait from unaffected parents.

b. Autosomal. A daughter inherited the trait from an unaffected father.

c. Both parents must be heterozygous for this child to have a $\frac{1}{4}$ chance of inheriting the disease. A's great-grandparents must both have been heterozygous, while B's grandparents must both have been heterozygous. A's paternal grandmother is unaffected, so she has a $\frac{2}{3}$ chance of being heterozygous; thus A's father has a $\frac{2}{3} \times \frac{1}{2} = \frac{1}{3}$ chance of being heterozygous, and A has a $\frac{2}{3} \times \frac{1}{2} \times \frac{1}{2} = \frac{1}{6}$ chance of being heterozygous. Similarly, B's father has a $\frac{2}{3}$ chance of being heterozygous, and B has a $\frac{2}{3} \times \frac{1}{2} = \frac{1}{3}$ chance of being heterozygous. Finally, the chance of an affected child is $\frac{1}{4} \times \frac{1}{6} \times \frac{1}{3} = \frac{1}{72}$.

26. H = hairy allele; h = smooth allele

plant 1: H/h
plant 2: H/H
plant 3: H/h
plant 4: h/h

Cross 2 indicates that hairy is dominant to smooth and both plant 1 and plant 3 are heterozygous. Plant 4 must be homozygous for the smooth allele since it exhibits the recessive trait, and plant 2 must be homozygous for the dominant allele since all progeny of plant 2 × plant 4 are hairy.

The progeny of cross 1 are $\frac{1}{2}$ H/H and $\frac{1}{2}$ H/h; of cross 2 are $\frac{1}{4}$ H/H, $\frac{1}{2}$ H/h, and $\frac{1}{4}$ h/h; of cross 3 are $\frac{1}{2}$ H/h and $\frac{1}{2}$ h/h; and of cross 4 all are H/h.

27.

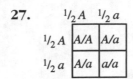

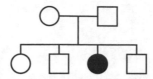

A pedigree is used to show genetic relationships between family members. Through deduction and interpretation of patterns, modes of inheritance may be indicated or ruled out. The genotypes of individuals represented in the pedigree can also be deduced in some cases. This information could prove useful both in understanding how genes affect phenotype and in providing genetic counseling. In a grid, a cross is represented as an abstract construct. It illustrates the probability of events that can happen but does not convey actual outcomes. Nonetheless, knowing that there is a $^{1}/_{4}$ chance of having a child that is *a/a* when both parents are *A/a* is useful information even if they already have four children that are not.

28. **a.** Pedigree 1: X-linked recessive

Affected individuals have inherited the disease from unaffected parents. This is a hallmark of recessive inheritance. The pedigree is also most consistent with the allele being X-linked for the following reasons: only males are affected; both affected progeny could have inherited the allele through their maternal parents who were either the daughter or grand-daughter of the original affected male; and you are told that the disease is rare so if it is not X-linked, both fathers of affected individuals that have married into this pedigree would also have to be carriers.

Pedigree 2: autosomal recessive

As in pedigree 1, the trait appears to be recessive. (This pedigree also indicates why it is more likely to see rare recessive traits in the progeny of consanguineous unions.) In this pedigree, it is more likely that the trait is autosomal. If the trait was X-linked, the affected son in the fourth generation would have inherited the trait from his mother yet his mother could not have inherited an X-linked trait from her unaffected father. Again, you are told that the disease is rare, so it is less likely that this affected son's grandmother was also a carrier.

b. Child of A and B:

Since this is an X-linked recessive trait, you must calculate B's probability of being a carrier; if she is, you must figure the probability of her passing that allele to her child and then multiply that by the probability that her child will be male.

B's grandmother must be a carrier. Thus, B's mother has a 50% chance of being a carrier and a 50% chance that if she is, she passed the allele to B. If B is a carrier, her children have a 50% chance of inheriting the allele and, if male, being affected. Therefore, A and B have a $\frac{1}{2} \times \frac{1}{2} \times \frac{1}{2} \times \frac{1}{2} = \frac{1}{16}$ chance of having an affected child.

Child of C and D:

This trait is inherited as an autosomal recessive; therefore, both C and D must be carriers in order for their child to have a 25% chance of being affected. Although D's probability can be calculated as $\frac{1}{2}$, since her mother must be heterozygous (a brother of D has an affected child), we do not know C's genotype. Since the trait is rare, we can assume that C has 0% chance of being a carrier and 0% chance of passing the allele to his child.

6

GENETIC RECOMBINATION IN EUKARYOTES

1. You are told that the two genes assort independently. Therefore,

From the first parent: $\frac{1}{4}$ the gametes will be A ; B

                                           $\frac{1}{4}$ the gametes will be A ; b

                                         $\frac{1}{4}$ the gametes will be a ; B

                                         $\frac{1}{4}$ the gametes will be a ; b

From the second parent: $\frac{1}{2}$ the gametes will be A ; b

                                         $\frac{1}{2}$ the gametes will be a ; b

and the progeny will be: $\frac{1}{8}$ A/A ; B/b

                                         $\frac{1}{8}$ A/A ; b/b

                                         $\frac{1}{4}$ A/a ; B/b

                                         $\frac{1}{4}$ A/a ; b/b

                                         $\frac{1}{8}$ a/a ; B/b

                                         $\frac{1}{8}$ a/a ; b/b

2. a., b. You are told that black is dominant to brown and intense is dominant to dilute. Thus, the brown and the dilute (recessive) traits must be homozygous to be expressed. Neither the black nor intense traits breed true so these must both be heterozygous.

Parents: dilute; Black × Intense; brown

d/d ; B/b D/d ; b/b

Progeny: dilute; Black d/d ; B/b

dilute; brown d/d ; b/b

Intense; Black D/d ; B/b

Intense; brown D/d ; b/b

3. a. 9 genotypes (just count)

b. 9:   1 R/R ; Y/Y 3:   1 R/R ; y/y 3:   1 r/r ; Y/Y 1:   1 r/r ; y/y

2 R/R ; Y/y 2 R/r ; y/y 2 r/r ; Y/y

4 R/r ; Y/y

2 R/r ; Y/Y

c. The number of different genotypes is 3^n, where n = number of genes. For simple dominant/recessive relationships, the number of different phenotypes is 2^n, where n = number of genes.

d. A round, yellow plant's genotype can be deduced either through a selfcross or testcross.

4. You perform the following cross and are told that the two genes are 10 m.u. apart.

$$A\ B/a\ b \times a\ b/a\ b$$

Among their progeny, 10% should be recombinant ($A\ b/a\ b$ and $a\ B/a\ b$) and 90% should be parental ($A\ B/a\ b$ and $a\ b/a\ b$). Therefore, $A\ B/a\ b$ should represent $1/2$ of the parentals, or 45%.

5. P $A\ d\,/\,A\ d \times a\ D\,/\,a\ D$

F$_1$ $A\ d\,/\,a\ D$

F$_2$ 1 $A\ d\,/\,A\ d$ Phenotype: A d

2 $A\ d\,/\,a\ D$ Phenotype: A D

1 $a\ D\,/\,a\ D$ Phenotype: a D

6. Since only parental types are recovered, the two genes must be tightly linked and recombination must be very rare. Knowing how many progeny were looked at would give an indication of how close the genes are.

7. The problem states that a female that is A/a . B/b is testcrossed. If the genes are unlinked, they should assort independently and the four progeny classes should be present in roughly equal proportions. This is clearly not the case. The A/a . B/b and a/a . b/b classes (the parentals) are much more common than the A/a . b/b and a/a . B/b classes (the recombinants). The two genes are on the same chromosome and are 10 map units apart.

$$RF = 100\% \times (46 + 54)/1000 = 10\%$$

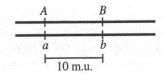

8. The cross is A/A . $b/b \times a/a$. B/B. The F_1 would be A/a . B/b.

a. If the genes are unlinked, all four progeny classes from the testcross (including a/a ; b/b) would equal 25%.

b. With completely linked genes, the F_1 would produce only $A\ b$ and $a\ B$ gametes. Thus, there would be a 0% chance of having $a\ b/a\ b$ progeny from a testcross of this F_1.

c. If the two genes are linked and 12 map units apart, 12% of the testcross progeny should be recombinants. Since the F_1 is $A\ b/a\ B$, $a\ b$ is one of the recombinant classes ($A\ B$ being the other) and should equal $^1/_2$ of the total recombinants or 6%.

d. 12% (see part **c**)

9. To answer this question, you must realize that (1) one chiasma involves two of the four chromatids of the homologous pair. So if 16% of the meioses have one chiasma, it will lead to 8% recombinants observed in the progeny (one half of the chromosomes of such a meiosis are still parental), and (2) half of the recombinants will be $B\ r$, so the correct answer is 4% (or b).

10. **a.** C/c ; $S/s \times C/c$; S/s There are 3 short:1 long, and 3 dark:1 albino.

b. C/C ; $S/s \times C/C$; s/s There are no albino, and there are 1 long:1 short.

c. C/c ; $S/S \times c/c$; S/S There are no long, and there are 1 dark:1 albino.

d. c/c ; $S/s \times c/c$; S/s All are albino, and there are 3 short:1 long.

e. C/c ; $s/s \times C/c$; s/s All are long, and there are 3 dark:1 albino.

f. C/C ; $S/s \times C/C$; S/s There are no albino, and there are 3 short:1 long.

g. C/c ; $S/s \times C/c$; s/s There are 3 dark:1 albino, and 1 short:1 long.

11. Cross 2 indicates that purple (G) is dominant to green (g), and cross 1 indicates cut (P) is dominant to potato (p).

Cross 1: G/g ; P/p × g/g ; P/p There are 3 cut:1 potato, and 1 purple: 1 green.

Cross 2: G/g ; P/p × G/g ; p/p There are 3 purple:1 green, and 1 cut: 1 potato.

Cross 3: G/G ; P/p × g/g ; P/p There are no green, and there are 3 cut: 1 potato.

Cross 4: G/g ; P/P × g/g ; p/p There are no potato, and there are 1 purple:1 green.

Cross 5: G/g ; p/p × g/g ; P/p There are 1 cut:1 potato, and there are 1 purple:1 green.

12. Assume there is no linkage. (This is your hypothesis. If it can be rejected, the genes are linked.) The expected values would be that genotypes occur with equal frequency. There are four genotypes in each case (n = 4) so there are 3 degrees of freedom.

$$\chi^2 = \Sigma \ (\text{observed-expected})^2/\text{expected}$$

Cross 1: $\chi^2 = [(310 - 300)^2 + (315 - 300)^2 + (287 - 300)^2 + (288 - 300)^2]/300$

$= 2.1266$; $p > 0.50$, nonsignificant; hypothesis cannot be rejected

Cross 2: $\chi^2 = [(36 - 30)^2 + (38 - 30)^2 + (23 - 30)^2 + (23 - 30)^2]/300$

$= 6.6$; $p > 0.10$, nonsignificant; hypothesis cannot be rejected

Cross 3: $\chi^2 = [(360 - 300)^2 + (380 - 300)^2 + (230 - 300)^2 + (230 - 300)^2]/300$

$= 66.0$; $p < 0.005$, significant; hypothesis must be rejected

Cross 4: $\chi^2 = [(74 - 60)^2 + (72 - 60)^2 + (50 - 60)^2 + (44 - 60)^2]/300$

$= 11.60$; $p < 0.01$, significant; hypothesis must be rejected

13. a. Note that only males are affected by both disorders. This suggests that both are X-linked recessive disorders. Using p for protan and P for non-protan and d for deutan and D for non-deutan, the inferred genotypes are listed on the pedigree below. The Y chromosome is shown, but the X is represented by the alleles carried.

b. Individual II-2 must have inherited both disorders in the transconfiguration (on separate chromosomes). Therefore, individual III-2 inherited both traits as the result of recombination (crossing-over) between his mother's X chromosomes.

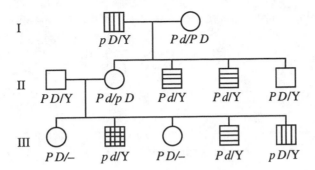

c. Since both genes are X-linked, this represents crossing-over. The progeny size is too small to give a reliable estimate of recombination.

14. **Unpacking the Problem**

a. There is no correct drawing; any will do. Pollen from the tassels is placed on the silks of the females. The seeds are the F_1 corn kernels.

b. The +'s all look the same because they signify wild type for each gene. The information is given in a specific order, which prevents confusion, at least initially. However, as you work the problem, which may require you to reorder the genes, errors can creep into your work if you do not make sure that you reorder the genes for each genotype in exactly the same way. You may find it easier to write the complete genotype, p^+, instead of +, to avoid confusion.

c. The phenotype is purple leaves and brown midriff to seeds. In other words, the two colors refer to different parts of the organism.

d. There is no significance in the original sequence of the data.

e. A tester is a homozygous recessive for all genes being studied. It is used so that the meiotic products in the organism being tested can be seen directly in the phenotype of the progeny.

f. The progeny phenotypes allow you to infer the genotypes of the plants. For example, *gre* stands for "green," the phenotype of $p^+/-$; *sen* stands for "virus-sensitive," the phenotype of $v^+/-$; and *pla* stands for "plain seed," the phenotype of $b^+/-$. In this test cross, all progeny have at least one recessive allele so the "gre sen pla" progeny are actually $p^+/p . v^+/v . b^+/b$.

g. *Gametes* refers to the gametes of the two pure-breeding parents. F_1 *gametes* refers to the gametes produced by the completely heterozygous F_1 progeny. They indicate whether crossing-over and independent assortment have occurred. In this case, because there is either independent assortment or crossing-over, or both, the data indicate that the three genes are not so tightly linked that zero recombination occurred.

h. The main focus is meiosis occurring in the F_1 parent.

i. The gametes from the tester are not shown because they contribute nothing to the phenotypic differences seen in the progeny.

j. Eight phenotypic classes are expected for three autosomal genes, whether or not they are linked, when all three genes have simple dominant-recessive relationships among their alleles. The general formula for the number of expected phenotypes is 2^n, where n is the number of genes being studied.

k. If the three genes were on separate chromosomes, the expectation is a 1:1:1:1:1:1:1:1 ratio.

l. The four classes of data correspond to the parentals (largest), two groups of single crossovers (intermediate), and double crossovers (smallest).

m. By comparing the parentals with the double crossovers, gene order can be determined. The gene in the middle "flips" with respect to the two flanking genes in the double-crossover progeny. In this case, one parental is +++ and one double crossover is p++. This indicates that the gene for leaf color (p) is in the middle.

n. If only two of the three genes are linked, the data can still be grouped, but the grouping will differ from that mentioned in (**l**) above. In this situation, the unlinked gene will show independent assortment with the two linked genes. There will be one class composed of four phenotypes in approximately equal frequency, which combined will total more than half the progeny. A second class will be composed of four phenotypes in approximately equal frequency, and the combined total will be less than half the progeny. For example, if the cross were $a\ b/+ +\ ;\ c/+ \times a\ b/a\ b\ ;\ c/c$, then the parental class (more frequent class) would have four components: $a\ b\ c$, $a\ b +$, $+ + c$, and $+ + +$. The recombinant class would be $a + c$, $a + +$, $+ b\ c$, and $+ b +$.

o. *Point* refers to locus. The usage does not imply linkage, but rather a testing for possible linkage. A four-point testcross would look like the following: $a/+\ .\ b/+\ .\ c/+\ .\ d/+ \times a/a\ .\ b/b\ .\ c/c\ .\ d/d$.

p. A *recombinant* refers to an individual who has alleles inherited from two different grandparents, both of whom were the parents of the individual's heterozygous parent. Another way to think about this term is that in the recombinant individual's heterozygous parent, recombination took place among the genes that were inherited from his or her parents. In this case, the recombination took place in the F_1 and the recombinants are among the F_2 progeny.

q. The "recombinant for" columns refer to specific gene pairs and progeny that exhibit recombination between those gene pairs.

r. There are three "recombinant for" columns because three genes can be grouped in three different gene pairs.

s. R refers to recombinant progeny, and they are determined by reference back to the parents of their heterozygous parent.

t. Column totals indicate the number of progeny that experience crossing-over between the specific gene pairs. They are used to calculate map units between the two genes.

u. The diagnostic test for linkage is a recombination frequency of less than 50%.

v. A map unit represents 1% crossing-over and is the same as a centimorgan.

w. In the tester, recombination cannot be detected in the gamete contribution to the progeny because the tester is homozygous. The F_1 individuals have genotypes fixed by their parents' homozygous state and, again, recombination cannot be detected in them, simply because their parents were homozygous.

x. Interference I = 1 − coefficient of coincidence = 1 − (observed double crossovers/expected double crossovers). The expected double crossovers are equal to p(frequency of crossing-over in the first region, in this case between v and p) × p(frequency of crossing-over in the second region, between p and b) × number of progeny. The probability of crossing-over is equal to map units converted back to percentage.

y. If the three genes are not all linked, then interference cannot be calculated.

z. A great deal of work is required to obtain 10,000 progeny in corn because each seed on a cob represents one progeny. Each cob may contain as many as 200 seeds. While seed characteristics can be assessed at the cob stage, for other characteristics, each seed must be separately planted and assessed after germination and growth. The bookkeeping task is also enormous.

Solution to the Problem

a. The three genes are linked.

b. Comparing the parentals (most frequent) with the double crossovers (least frequent), the gene order is $v\,p\,b$. There were 2200 recombinants between v and p, and 1500 between p and b. The general formula for map units is

m.u. = 100%(number of recombinants)/total number of progeny

Therefore, the map units between v and p = 100%(2200)/10,000 = 22 m.u., and the map units between p and b = 100%(1500)/10,000 = 15 m.u.

The map is

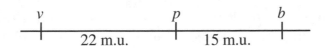

c. I = 1 − observed double crossovers/expected double crossovers

= 1 − 132/(0.22)(0.15)(10,000)

= 1 − 0.4 = 0.6

15. a. By comparing the two most frequent classes (parentals: $an\;br^+\,f^+$, $an^+br\,f$) to the least frequent classes (DCO: $an^+\;br\,f^+$, $an\;br^+\,f$), the gene order can be determined. The gene in the middle switches with respect to the other two (the order is $an\,f\,br$). Now the crosses can be written fully:

P $an\,f^+\,br^+/an\,f^+\,br^+$ × $an^+\,f\,br/an^+\,f\,br$

F_1 $an\,f^+\,br^+/an^+\,f\,br$ × $an\,f\,br/an\,f\,br$

F_2 355 $an\,f^+\,br^+/an\,f\,br$ Parental

339    $an^+\,f\,br/an\,f\,br$ Parental

88    $an^+\,f^+\,br^+/an\,f\,br$ CO an—f

55    $an\,f\,br/an\,f\,br$ CO an—f

21    $an^+\,f\,br^+/an\,f\,br$ CO f—br

17    $an\,f^+\,br/an\,f\,br$ CO f—br

2    $an^+\,f^+\,br/an\,f\,br$ DCO

2    $an\,f\,br^+/an\,f\,br$ DCO

b. *an—f*: 100%(88 + 55 + 2 + 2)/879 = 16.72 m.u.

f—br: 100%(21 + 17 + 2 + 2)/879 = 4.78 m.u.

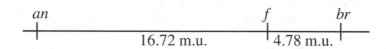

c. Interference = 1 – (observed DCO/expected DCO)

= 1 – 4/(0.1672)(0.0478)(879)

= 1 – 0.569 = 0.431

16. Dear Brother Mendel:

I have recently read your most engrossing manuscript detailing the results of your most wise experiments with garden peas. I salute both your curiosity and your ingenuity in conducting said experiments, thereby opening up for scientific exploration an entire new area of our Maker's universe. Dear Sir, your findings are extraordinary!

While I do not pretend to compare myself to you in any fashion, I beg to bring to your attention certain findings I have made with the aid of that most fascinating and revealing instrument, the microscope. I have been turning my attention to the smallest of worlds with an instrument that I myself have built, and I have noticed some structures that may parallel in behavior the factors that you have postulated in the pea.

I have worked with grasshoppers, however, not your garden peas. Although you are a man of the cloth, you are also a man of science, and I pray that you will not be offended when I state that I have specifically studied the reproductive organs of male grasshoppers. Indeed, I did not limit myself to studying the organs themselves; instead, I also studied the smaller units that make up the male organs and have beheld structures most amazing within them.

These structures are contained within numerous small bags within the male organs. Each bag has a number of these structures, which are long and threadlike at some times and short and compact at other times. They come together in the middle of a bag, and then they appear to divide equally. Shortly thereafter, the bag itself divides, and what looks like half of the threadlike structures goes into each new bag. Could it be, Sir, that these threadlike structures are the very same as your factors? I know, of course, that garden peas do not have male organs in the same way that grasshoppers do, but it seems to me that you found it necessary to emasculate the garden peas in order to do some crosses, so I do not think it too far-fetched to postulate a similarity between grasshoppers and garden peas in this respect.

Pray, Sir, do not laugh at me and dismiss my thoughts on this subject even though I have neither your excellent training nor your astounding wisdom in the Sciences. I remain your humble servant to eternity!

17. Two unlinked genes; "no dots" (D) is dominant to "dots" (d) and "vertical bar" (H) is dominant to "horizontal bar" (h).

dots, Vertical bar $\times$ No dots, horizontal bar

d/d ; H/H $\qquad$ D/D ; h/h

$\downarrow$

All: No dots, Vertical bar

$\qquad$ D/d ; H/h $\times$ dots, horizontal bar

$\qquad\qquad\qquad$ d/d ; h/h

$\qquad\qquad$ $\downarrow$

$\qquad$ $1/4$ $\quad$ dots, horizontal bar

$\qquad\qquad\qquad$ d/d ; h/h

$\qquad$ $1/4$ $\quad$ No dots, horizontal bar

$\qquad\qquad\qquad$ D/d ; h/h

$\qquad$ $1/4$ $\quad$ dots, Vertical bar

$\qquad\qquad\qquad$ d/d ; H/h

$\qquad$ $1/4$ $\quad$ No dots, Vertical bar

$\qquad\qquad\qquad$ D/d ; H/h

18. Two unlinked genes; "no dots" (*d*) is recessive to "dots" (*d*⁺) and "no lines" (*l*) is recessive to "lines" (*l*⁺).

no dots, Lines   ×   Dots, no lines

d/d ; l^+/l^+ d^+/d^+ ; l/l

↓

All: Dots, Lines

d^+/d ; l^+/l

↓

Self: $^9/_{16}$ Dots, Lines

$d^+/-$; $l^+/-$

$^3/_{16}$ Dots; no lines

$d^+/-$; l/l

$^3/_{16}$ no dots, Lines

d/d ; $l^+/-$

$^1/_{16}$ no dots, no lines

d/d ; l/l

19. Two linked genes, 20 m.u. apart; "no dots" (*d*) is recessive to "dots" (*D*) and "no line" (*l*) is recessive to "line" (*L*).

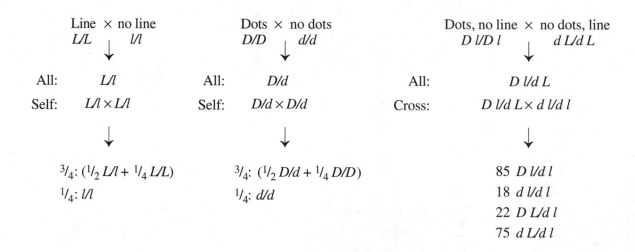

Line × no line
L/L ↓ l/l

All: L/l

Self: $L/l \times L/l$

↓

$^3/_4$: ($^1/_2\ L/l + ^1/_4\ L/L$)
$^1/_4$: l/l

Dots × no dots
D/D ↓ d/d

All: D/d

Self: $D/d \times D/d$

↓

$^3/_4$: ($^1/_2\ D/d + ^1/_4\ D/D$)
$^1/_4$: d/d

Dots, no line × no dots, line
$D\ l/D\ l$ ↓ $d\ L/d\ L$

All: $D\ l/d\ L$

Cross: $D\ l/d\ L \times d\ l/d\ l$

↓

85 $D\ l/d\ l$
18 $d\ l/d\ l$
22 $D\ L/d\ l$
75 $d\ L/d\ l$

20. Two X-linked genes, 10 m.u. apart; "white" (b) is recessive to "black" (b^+) and "wavy tail" (s) is recessive to "straight tail" (s^+).

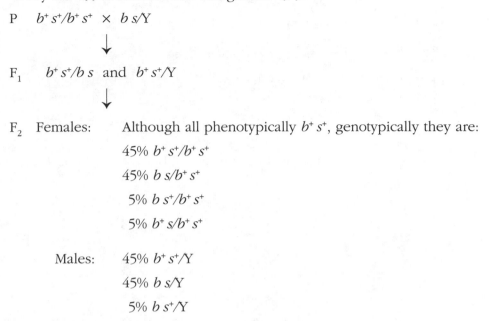

P $b^+ s^+/b^+ s^+ \times b s/Y$

F$_1$ $b^+ s^+/b s$ and $b^+ s^+/Y$

F$_2$ Females: Although all phenotypically $b^+ s^+$, genotypically they are:

    45% $b^+ s^+/b^+ s^+$

    45% $b s/b^+ s^+$

     5% $b s^+/b^+ s^+$

     5% $b^+ s/b^+ s^+$

   Males: 45% $b^+ s^+/Y$

    45% $b s/Y$

     5% $b s^+/Y$

     5% $b^+ s/Y$

21. Two linked genes, 30 m.u. apart. (Remember, fungi are haploid.)

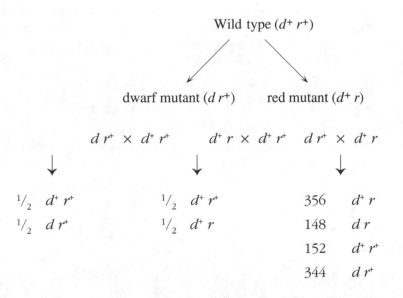

Wild type ($d^+ r^+$)

dwarf mutant ($d r^+$)   red mutant ($d^+ r$)

    $d r^+ \times d^+ r^+$ $d^+ r \times d^+ r^+$ $d r^+ \times d^+ r$

$\frac{1}{2}$ $d^+ r^+$ $\frac{1}{2}$ $d^+ r^+$ 356 $d^+ r$

$\frac{1}{2}$ $d r^+$ $\frac{1}{2}$ $d^+ r$ 148 $d r$

                                       152    $d^+ r^+$

                                       344    $d r^+$

22. **a.** Since each gene assorts independently, each probability should be considered separately and then all multiplied together for the answer.

For (1) $\frac{1}{2}$ will be A, $\frac{3}{4}$ will be B, $\frac{1}{2}$ will be C, $\frac{3}{4}$ will be D, and $\frac{1}{2}$ will be E.

$$\tfrac{1}{2} \times \tfrac{3}{4} \times \tfrac{1}{2} \times \tfrac{3}{4} \times \tfrac{1}{2} = \tfrac{9}{128}$$

For (2) $\frac{1}{2}$ will be a, $\frac{3}{4}$ will be B, $\frac{1}{2}$ will be c, $\frac{3}{4}$ will be D, and $\frac{1}{2}$ will be e.

$$\frac{1}{2} \times \frac{3}{4} \times \frac{1}{2} \times \frac{3}{4} \times \frac{1}{2} = \frac{9}{128}$$

For (3) it is the sum of (1) and (2) = $\frac{9}{128} + \frac{9}{128} = \frac{9}{64}$

For (4) it is $1 - (\text{part } 3) = 1 - \frac{9}{64} = \frac{55}{64}$

b. For (1) $\frac{1}{2}$ will be A/a, $\frac{1}{2}$ will be B/b, $\frac{1}{2}$ will be C/c, $\frac{1}{2}$ will be D/d, and $\frac{1}{2}$ will be E/e.

$$\frac{1}{2} \times \frac{1}{2} \times \frac{1}{2} \times \frac{1}{2} \times \frac{1}{2} = \frac{1}{32}$$

For (2) $\frac{1}{2}$ will be a/a, $\frac{1}{2}$ will be B/b, $\frac{1}{2}$ will be c/c, $\frac{1}{2}$ will be D/d, and $\frac{1}{2}$ will be e/e.

$$\frac{1}{2} \times \frac{1}{2} \times \frac{1}{2} \times \frac{1}{2} \times \frac{1}{2} = \frac{1}{32}$$

For (3) it is the sum of (1) and (2) = $\frac{1}{16}$.

For (4) it is $1 - (\text{part } 3) = 1 - \frac{1}{16} = \frac{15}{16}$

23. The data given for each of the three-point testcrosses can be used to determine the gene order by realizing that the rarest recombinant classes are the result of double cross-over events. Compare these chromosomes to the "parental" types—the alleles that have switched represent the gene in the middle.

For example, in (1), the most common phenotypes (+ + + and a b c) represent the parental allele combinations. Comparing these to the rarest phenotypes of this data set (+ b c and a + +) indicates that the *a* gene is recombinant and must be in the middle. The gene order is *b a c*.

For (2), + b c and a + + (the parentals) should be compared to + + + and a b c (the rarest recombinants) to indicate that the *a* gene is in the middle. The gene order is *b a c*.

For (3), compare + b + and a + c with a b + and + + c, which gives the gene order *b a c*.

For (4), compare + + c and a b + with + + + and a b c, which gives the gene order *a c b*.

For (5), compare + + + and a b c with + + c and a b +, which gives the gene order *a c b*.

24. **a.** The first F$_1$ is *L H/l h* and the second is *l H/L h*. For progeny that are *l h/l h*, they have received a "parental" chromosome from the first F$_1$ and a "recom-

binant" chromosome from the second F_1. The genes are 16% apart so the chance of a parental chromosome is $\frac{1}{2}(100 - 16\%) = 42\%$ and the chance of a recombinant chromosome is $\frac{1}{2}(16\%) = 8\%$.

$$\text{The chance of both events} = 42\% \times 8\% = 3.36\%$$

b. To obtain *Lb/l b* progeny, either a parental chromosome from each parent was inherited *or* a recombinant chromosome from each parent was inherited. The total probability will therefore be

$$(42\% \times 42\%) + (8\% \times 8\%) = (17.6\% + 0.6\%) = 18.2\%.$$

25. Since there is no branch migration, the heteroduplex that occurs is only the result of strand invasion. All heteroduplexes are repaired, but there is a bias to repair the mismatch to the *A* allele over the *a* allele (80% to *A* but only 20% to *a*).

Strand invasion by *A* would result in a transient heteroduplex (mismatch) that would be repaired to *A* 80% of the time and result in an aberrant 6:2 ratio (*A:a*). In the other 20% of these cases, the *A* would be converted to an *a*, resulting in the normally expected 4:4 ratio. Strand invasion by *a* would result in a transient heteroduplex that would be repaired to *a* 20% of the time and result in an aberrant 2:6 ratio (*A:a*). In the other 80% of these cases, the *a* would be converted to an *A*, resulting in the normally expected 4:4 ratio. Overall then, 80% of the aberrant asci will show a 6:2 ratio and 20% will show a 2:6 ratio. Since all heteroduplexes are repaired, no 5:3 nor 3:5 aberrant ratios will be observed.

26. (1) Impossible: The alleles *A* and *a* should be on homologous chromosomes as should the alleles *B* and *b*.

(2) Meiosis II: Sister chromatids are separating, and there is only one copy of each chromosome.

(3) Meiosis II: Same as (2).

(4) Meiosis II: Same as (2).

(5) Mitosis: Sister chromatids are separating, and there are two copies of each chromosome.

(6) Impossible: Sister chromatids are nonidentical for all chromosomes.

(7) Impossible: There are 4 copies of each chromosome.

(8) Impossible: Same as (7).

(9) Impossible: Same as (7).

(10) Meiosis I: Homologous chromosomes are separating.

(11) Impossible: All four chromatids of each homologous chromosome have the same allele.

(12) Impossible: Same as (1).

27. The rare prototrophic colonies could be the result of intragenic recombination between the two mutations of the *hist-1* gene. Since both mutations *never* revert, assume that both are the result of small deletions. Reciprocal recombination between these will result in one wild-type allele and one doubly mutant allele (which would not appear different than either single mutation). Only the wild-type alleles would be observed and these would represent one half of the total recombinants.

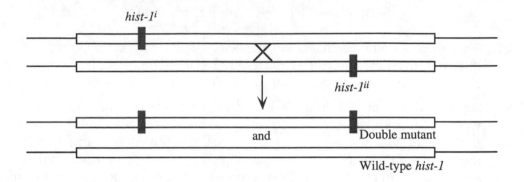

28. **a.** Long (s^+) is dominant to short (s), since the second cross has short progeny from long parents. Large (v^+) is dominant to vestigial (v), since the third cross has vestigial progeny from large parents.

    b. Since short and vestigial are recessive, only flies homozygous for the respective alleles will express these traits.

Cross 1:	v^+/v ; s^+/s^+ × v/v ; s/s	Long parent is true-breeding, large is not.
Cross 2:	v^+/v ; s^+/s × v/v ; s^+/s	Large parent not true-breeding; 3 long:1 short.
Cross 3:	v^+/v ; s^+/s × v^+/v ; s/s	Long parent not true-breeding; 3 large:1 vestigial.
Cross 4:	v^+/v ; s/s × v/v ; s^+/s	Neither large nor long parents are true-breeding.

29. **a., b., c.** Since all eight allelic combinations are equally likely in the gametes, it can be inferred that the three genes are on separate chromosomes. The following figures use chromosome size and centromere placement to distinguish the three nonhomologous chromosomes.

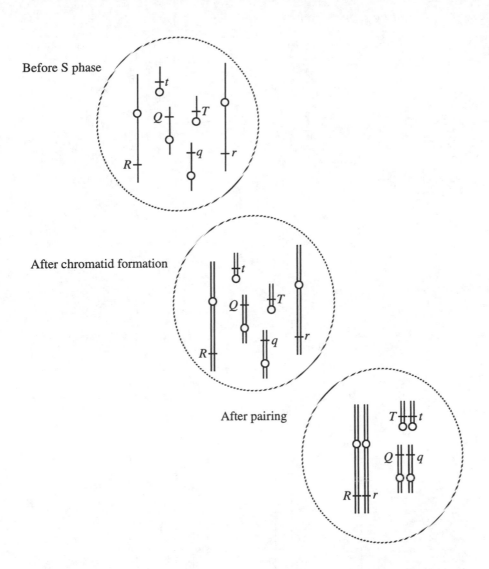

Before S phase

After chromatid formation

After pairing

d. and e. Independent assortment of nonhomologous chromosomes will give 8 possible allelic combinations. This is graphically represented by showing the various alignments the separate chromosomes may take during meiosis. The segregation of homologous chromosomes during anaphase I and then the splitting of sister chromatids during anaphase II is schematically indicated. Although it is very likely that crossing-over will take place during prophase I, it will not affect the genotypic ratios of the gametes and thus is ignored.

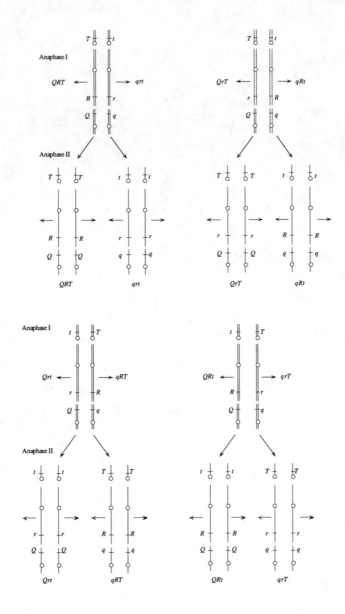

30. **a.** From the data, it can be concluded that the genes for flower color and plant height are linked (on the same chromosome). The position of the gene for leaf width cannot be determined from this data since all progeny express the dominant trait. In the tall, red parent, the alleles for tall and white are in the cis configuration (on the same chromosome) as are the alleles for short and red. Thus, the progeny that are tall, white or short, red represent the parental chromosomes and the tall, red or short, white are recombinants. The two genes are 100%(21 + 19)/total = 4 map units apart.

Tall, Red, Wide × short, white, narrow

$$\frac{T \quad r}{t \quad R} \cdot \frac{W}{W} \times \frac{t \quad r}{t \quad r} \cdot \frac{w}{w}$$

b. The chance of obtaining short, white, wide progeny = $p(t\,r)^2 = (^{4\%}/_2)^2 = 0.04\%$.

7 RECOMBINATION IN BACTERIA AND THEIR VIRUSES

1. An Hfr strain has the fertility factor F integrated into the chromosome. An F+ strain has the fertility factor free in the cytoplasm. An F⁻ strain lacks the fertility factor.

2. All cultures of F⁺ strains have a small proportion of cells in which the F factor is integrated into the bacterial chromosome and are, by definition, Hfr cells. These Hfr cells transfer markers from the host chromosome to a recipient during conjugation.

3. a. Hfr cells involved in conjugation transfer host genes in a linear fashion. The genes transferred depend on both the Hfr strain and the length of time during which the transfer occurs. Therefore, a population containing several different Hfr strains will appear to have an almost random transfer of host genes. This is similar to generalized transduction, in which the viral protein coat forms around a specific amount of DNA rather than specific genes. In generalized transduction, any gene can be transferred.

b. F′ factors arise from improper excision of an Hfr from the bacterial chromosome. They can have only specific bacterial genes on them because the integration site is fixed for each strain. Specialized transduction resembles this in that the viral particle integrates into a specific region of the bacterial chromosome and then, upon improper excision, can take with it only specific bacterial genes. In both cases, the transferred gene exists as a second copy.

4. Generalized transduction occurs with lytic phages that enter a bacterial cell, fragment the bacterial chromosome, and then, while new viral particles are being assembled, improperly incorporate some bacterial DNA within the viral protein coat. Because the amount of DNA, not the information content of the DNA, is what governs viral particle formation, any bacterial gene can be included within the newly formed virus. In contrast, specialized transduction occurs with improper excision of viral DNA from the host chromosome in lysogenic phages. Because the integration site is fixed, only those bacterial genes very close to the integration site will be included in a newly formed virus.

5. While the interrupted-mating experiments will yield the gene order, it will be relative only to fairly distant markers. Thus, the precise location cannot be pinpointed with this technique. Generalized transduction will yield information with regard to very close markers, which makes it a poor choice for the initial experiments because of the massive amount of screening that would have to be done. Together, the two techniques allow, first, for a localization of the mutant (interrupted-mating) and, second, for precise determination of the location of the mutant (generalized transduction) within the general region.

6. This problem is analogous to forming long gene maps with a series of three-point testcrosses. Arrange the four sequences so that their regions of overlap are aligned:

$$\overline{M-Z}-X-W-C$$
$$W-C-N-A-L$$
$$A-L-B-R-U$$
$$B-R-U-\overline{M-Z}$$

The regions with the bars above or below are identical in sequence (and "close" the circular chromosome). The correct order of markers on this circular map is

$$-M-Z-X-W-C-N-A-L-B-R-U-$$

7. To interpret the data, the following results are expected:

Cross	Result
F$^+$ × F$^-$	(L) low number of recombinants
Hfr × F$^-$	(M) many recombinants
Hfr × Hfr	(0) no recombinants
Hfr × F$^+$	(0) no recombinants
F$^+$ × F$^+$	(0) no recombinants
F$^-$ × F$^-$	(0) no recombinants

The only strains that show both the (L) and the (M) result when crossed are 2, 3, and 7. These must be F⁻ since that is the only cell type that can participate in a cross and give either recombination result. Hfr strains will result in only (M) or (0), and F⁺ will result in only (L) or (0) when crossed. Thus, strains 1 and 8 are F⁺, and strains 4, 5, and 6 are Hfr.

8. Prototrophic strains of *E. coli* will grow on minimal media while auxotrophic strains will only grow on media supplemented with the required molecule(s). Thus, strain 3 is prototrophic (wild-type), strain 4 is *met⁻*, strain 1 is *arg⁻*, and strain 2 is *arg⁻ met⁻*.

9. a.

Agar type	Selected genes
1	c^+
2	a^+
3	b^+

b. The order of genes is revealed in the sequence of colony appearance. Because colonies first appear on agar type 1, which selects for c^+, c must be first. Colonies next appear on agar type 3, which selects for b^+, indicating that b follows c. Allele a^+ appears last. The gene order is $c\ b\ a$. The three genes are roughly equally spaced.

c. In this problem you are looking at the cotransfer of the selected gene with the d^- allele (both from the Hfr). Cells that are d^- do not grow because the medium is lacking D and selecting for those cells that are d^+. Therefore, the farther a gene is from gene d, the less likely cotransfer of the selected gene will occur with d^- and the more likely that colonies will grow (remain d^+). From the data, d is closest to b (only $8/100$ did not cotransfer d^- with b^+.) It is also closer to a than it is to c. Thus the gene order is $c\ b\ d\ a$ (or $a\ d\ b\ c$).

d. With no A or B in the agar, the medium selects for $a^+\ b^+$, and the first colonies should appear at about 17.5 minutes.

10. *Unpacking the Problem*

a. *E. coli* is a bacterium and a prokaryote.

b. *E. coli* can be grown in suspension or on an agar medium. The latter method allows for the identification of individual colonies, each a clone of descendants from a single cell (and visible to the naked eye when it reaches more than 10^7 cells).

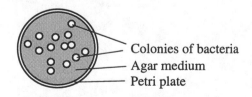

c. Naturally, *E. coli* is an enteric bacterium living symbiotically within the gut of host organisms (like us).

d. Minimal medium consists of inorganic salts, a carbon source for energy, and water.

e. Prototroph refers to the wild-type phenotype, or in other words, an organism that can grow on minimal media. Auxotroph refers to a mutant that can grow only on a medium supplemented with one or more specific nutrients not required by the wild-type strain.

f. In this experiment, the Hfr and the exconjugants that can grow on minimal medium are prototrophs, whereas the recipient F⁻ and the exconjugants that do not grow on minimal medium are auxotrophs.

g. Unknown strains would be grown as individual colonies on medium enriched with proline and thiamine, and then cells from each colony could be picked (by a sterile toothpick, for example) and placed individually onto medium supplemented with either thiamine or proline or onto minimal medium. Proline and thiamine auxotrophs would be identified on the basis of growth patterns. For example, a *pro⁻* strain will grow only on medium supplemented with proline.

Instead of the labor-intensive method of individually picking cells, replica plating can be used to transfer some cells of each colony from a master plate (supplemented with proline and thiamine) to plates that contain the various media described above. The physical arrangement (and positional patterns) of colonies is used to identify the various colonies as they are transferred from plate to plate.

h. Proline is an amino acid and thiamine is a B_1 vitamin. Their chemical nature does not matter to the experiment other than that they are necessary chemicals for cell growth that prototrophs can synthesize from ingredients in minimal medium and specific auxotrophic mutants cannot.

i.

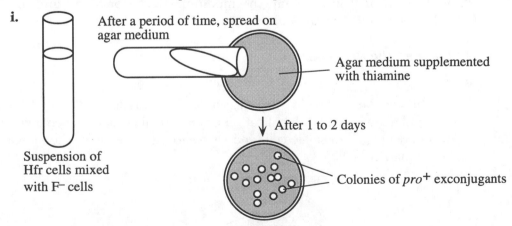

After a period of time, spread on agar medium

Agar medium supplemented with thiamine

After 1 to 2 days

Suspension of Hfr cells mixed with F⁻ cells

Colonies of *pro⁺* exconjugants

j. Interrupted-mating experiments are used to roughly map genes onto the circular bacterial chromosome.

k. The Hfr and F⁻ strains are mixed together in solution, and then at various times, samples are removed and put into a kitchen blender, vortexed (blender is turned on) for a few seconds to disrupt conjugation, and then plated onto a medium containing the appropriate supplements. The amount of time that has passed from the mixing of the strains to mating disruption is used as a measurement for mapping. The time of first appearance of a specific gene from the Hfr in the F⁻ cell gives the gene's relative position in minutes. Typically, the F⁻ cells are streptomycin-resistant and the Hfr cells are streptomycin-sensitive. The antibiotic is used in the various media to kill the Hfr cells (which are otherwise prototrophic) and allow only those F⁻ cells that have received the appropriate gene or genes from the Hfr to grow. In this case, it would be discovered that some of the F⁻ cells would become *thi*⁺ in samples taken earlier in the experiment than samples taken when they first become *pro*⁺.

l. In this experiment, there is no attempt to disrupt conjugation. The two strains are mixed and at some later (unspecified) time, plated onto medium containing thiamine. This selects for strains that are *pro*⁺, since proline is not present in this medium.

m. Exconjugants are recipient cells (F⁻) that now contain alleles from the donor (Hfr). Typically, the F⁻ cells are streptomycin-resistant and the Hfr cells are streptomycin-sensitive. The antibiotic is used in the various media to kill the Hfr cells and allow only the appropriate F⁻ exconjugants to grow.

n. The statement "*pro* enters after *thi*" is one of gene position and order relative to the transfer of the bacterial chromosome by a particular Hfr. For the Hfr in this experiment, transfer occurs such that the *pro* gene is transferred after the *thi* gene. Since this Hfr is also *pro*⁺, it is this specific allele that is entering.

o. In this experiment, "fully supplemented" medium contains proline and thiamine.

p. All exconjugants are *pro*⁺, since that is the way they were selected. Thus, those that do not grow on minimal medium must require thiamine.

q. Genetic exchange in prokaryotes does not take place between two whole genomes as it does in eukaryotes. It takes place between one complete circular genome, the F⁻, and an incomplete linear genomic fragment donated by the Hfr. In this way, exchange of genetic information is nonreciprocal (from Hfr to F⁻). Only even numbers of crossovers are allowed between the two DNAs, since the circular chromosome would become linear otherwise. This results in unidirectional exchange, since part of the DNA of the recipient chromosome is replaced by the DNA of the donor, while the other product (the rest of the donor DNA now with some recombined recipient DNA) is nonviable and lost.

r. In this experiment, the map distance is calculated by selecting for the last marker to enter (in this case *pro⁺*) and then determining how often the earlier unselected marker (in this case *thi⁺*) is also present. Look at the following diagram:

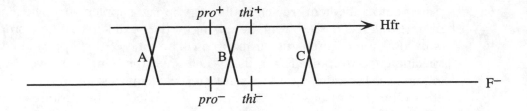

For the F⁻ cell to become *pro⁺*, two recombination events have to occur—one in the region to the left (marked A) and one in either region to the right (marked B or C.) Thus the percentage of *pro⁺* (second recombination within either B or C) that are *thi⁻* (second event only within region B) can be used to determine map distance where 1% = 1 map unit.

Solution to the Problem

a. The two genotypes being cultured are *pro⁺ thi⁻* (grows only on media supplemented with thiamine) and *pro⁺ thi⁺* (grows on minimal media.)

b. Two recombination events must occur, one on either side of *pro* (since exconjugants were plated on medium supplemented with thiamine, only *pro⁺* cells would have grown). The *pro⁺ thi⁻* strains would have had recombination in regions labeled A and B, and the *pro⁺ thi⁺* strains would have had recombination in regions labeled A and C.

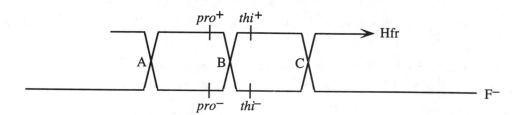

c. The distance between *pro* and *thi* is =

$$\frac{100\%(\text{the number of colonies that are } pro^+ \, thi^-)}{\text{total number of colonies}}$$

$$= 100\%(40)/360 = 11.1\%$$

11. **a.** Determine the gene order by comparing *arg⁺ bio⁺ leu⁻* with *arg⁺ bio⁻ leu⁺*. If the order were *arg leu bio*, four crossovers would be required to get *arg⁺ leu⁻ bio⁺*, while only two would be required to get *arg⁺ leu⁺ bio⁻*. If the order is *arg bio leu*, four crossovers would be required to get *arg⁺ bio⁻ leu⁺*, and only two would be required to get *arg⁺ bio⁺ leu⁻*. There are eight recombinants that are *arg⁺ bio⁺ leu⁻* and none that are *arg⁺ bio⁻ leu⁺*. On the basis of the frequencies of these two classes, the gene order is *arg bio leu*.

   b. The *arg-bio* distance is determined by calculating the percentage of the exconjugants that are *arg⁺ bio⁻ leu⁻*. These cells would have had a crossing-over event between the *arg* and *bio* genes.

$$RF = 100\%(48)/376 = 12.76 \text{ m.u.}$$

   Similarly, the *bio-leu* distance is estimated by the *arg⁺ bio⁺ leu⁻* colony type.

$$RF = 100\%(8)/376 = 2.12 \text{ m.u.}$$

12. The best explanation is that the integrated F factor of the Hfr looped out of the bacterial chromosome abnormally and is now an F′ that contains the *pro⁺* gene. This F′ is rapidly transferred to F⁻ cells, converting them to *pro⁺* (and F⁺).

13. The high rate of integration and the preference for the same site originally occupied by the F factor suggest that the F′ contains some homology with the original site. The source of homology could be a fragment of the F factor, or more likely, it is homology with the chromosomal copy of the bacterial gene that is also present on the F′.

14. First carry out a cross between the Hfr and F⁻, and then select for colonies that are *ala⁺ str⁻*. If the Hfr donates the *ala* region late, then redo the cross but now interrupt the mating early and select for *ala⁺*. This selects for an F′, since this Hfr would not have transferred the *ala* gene early.

    If the Hfr instead donates this region early, then use a Rec⁻ strain that cannot incorporate a fragment of the donor chromosome by recombination. Any *ala⁺* colonies from the cross should then be used in a second mating to another *ala⁻* strain to see whether they can donate the *ala* gene easily, which would indicate that there is F′ *ala*. (This would also require another marker to differentiate the donor and recipient strains. For example, the *ala⁻* strain could be tetracycline⁻ and selection would be for *ala⁺ tet⁻*.)

15. a., b.

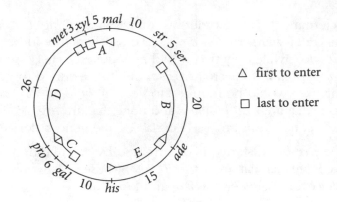

△ first to enter

□ last to enter

c. A: Select for *mal⁺*

B: Select for *ade⁺*

C: Select for *pro⁺*

D: Select for *pro⁺*

E: Select for *his⁺*

16. a. If the two genes are far enough apart to be located on separate DNA fragments, then the frequency of double transformants should be the product of the frequency of the two single transformants, or (4.3%) × (0.40%) = 0.017%. The observed double transformant frequency is 0.17%, a factor of 10 greater than expected. Therefore, the two genes are located close enough together to be cotransformed at a rate of 0.17%.

b. Here, when the two genes must be contained on separate pieces of DNA, the rate of cotransformation is much lower, confirming the conclusion in part **a.**

17. No. Closely linked loci would be expected to be cotransduced; the greater the cotransduction frequency, the closer the loci are. Since only 1 of 858 *metE⁺* was also *pyrD⁺*, the genes are not closely linked. The lone *metE⁺ pyrD⁺* could be the result of cotransduction, or it could be a spontaneous mutation of *pyrD* to *pyrD⁺*, or the result of co-infection by two separate transducing phage.

18. a. This is calculated as the percentage of *pur⁺* colonies that are also *nad⁺*:

$$= 100\%(3 + 10)/50 = 26\%$$

b. This is calculated as the percentage of *pur⁺* colonies that are also *pdx⁻*:

$$= 100\%(10 + 13)/50 = 46\%$$

c. *pdx* is closer, as determined by cotransduction rates.

d. From the cotransduction frequencies, you know that *pdx* is closer to *pur* than *nad* is, so there are two gene orders possible: *pur pdx nad* or *pdx pur nad*. Now, consider how a bacterial chromosome that is *pur⁺ pdx⁺ nad⁺*

might be generated given the two gene orders: if *pdx* is in the middle, 4 crossovers are required to get *pur⁺ pdx⁺ nad⁺*; if *pur* is in the middle, only 2 crossovers are required (see below). The results indicate that there are fewer *pur⁺ pdx⁺ nad⁺* transductants than any other class suggesting that this class is "harder" to generate than the others. This implies that *pdx* is in middle and the gene order is *pur pdx nad*.

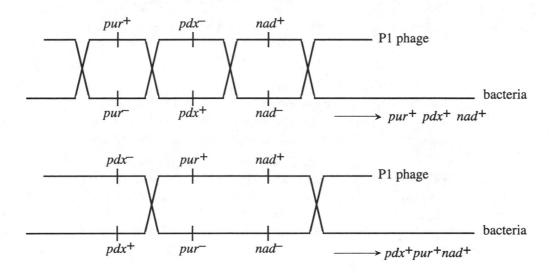

19. **a.** The parental genotypes are + + + and *m r tu*. For determining the *m-r* distance, the recombinant progeny are

m + tu	162
m + +	520
+ r tu	474
+ r +	172
	1328

Therefore the map distance is 100%(1328)/10,342 = 12.8 m.u.

Using the same approach, the *r-tu* distance is 100%(2152)/10,342 = 20.8 m.u., and the *m-tu* distance is 100%(2812)/10,342 = 27.2 m.u.

b. Since genes *m* and *tu* are the farthest apart, the gene order must be *m r tu*.

c. The coefficient of coincidence (c.o.c.) compares the actual number of double crossovers to the expected number, (where c.o.c. = observed double crossovers/expected double crossovers). For these data, the expected number of double recombinants is (0.128)(0.208)(10,342) = 275. Thus, c.o.c. = (162 + 172)/275 = 1.2. This indicates that there are more double crossover events than predicted and suggests that the occurrence of one crossover makes a second crossover between the same DNA molecules more likely to occur.

20. The most straightforward way would be to pick two Hfr strains that are near the genes in question but are oriented in opposite directions. Then, measure the time of transfer between two specific genes, in one case when they are transferred early and in the other when they are transferred late. For example,

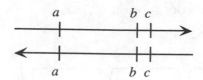

21. a. To determine which genes are close, compare the frequency of double transformants. Pairwise testing gives low values whenever B is involved but fairly high rates when any drug but B is involved. This suggests that the gene for B resistance is not close to the other three genes.

b. To determine the relative order of genes for resistance to A, C, and D, compare the frequency of double and triple transformants. The frequency of resistance to AC is approximately the same as resistance to ACD. This strongly suggests that D is in the middle. Also, the frequency of AD co-resistance is higher than AC (suggesting that the gene for A resistance is closer to D than to C), and the frequency of CD is higher than AC (suggesting that C is closer to D than to A).

22. In a small percent of the cases, *gal*⁺ transductants can arise by recombination between the *gal*⁺ DNA of the ldgal transducing phage and the *gal*⁻ gene on the chromosome. This will generate *gal*⁺ transductants without phage integration.

23. a. This appears to be specialized transduction. It is characterized by the transduction of specific markers based on the position of the integration of the prophage. Only those genes near the integration site are possible candidates for misincorporation into phage particles that then deliver this DNA to recipient bacteria.

b. The only media that supported colony growth were those lacking either cysteine or leucine. These selected for *cys*⁺ or *leu*⁺ transductants and indicate that the prophage is located in the *cys-leu* region.

24. a., b.

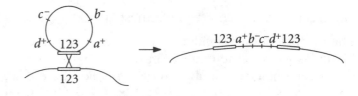

c.

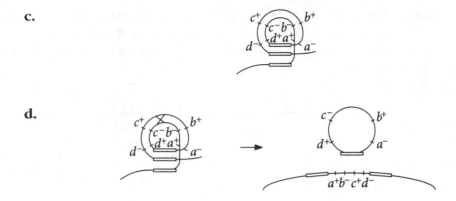

d.

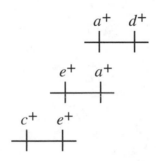

25. If a compound is not added and growth occurs, the *E. coli* recipient cell must have received the wild-type genes for production of those nutrients by transduction. Thus, the BCE culture selects for cells that are now a^+ and d^+, the BCD culture selects for cells that are a^+ and e^+, and the ABD culture selects for cells that are c^+ and e^+. These genes can be aligned (see below) to give the map order of *d a e c*. (Notice that *b* is never cotransduced and is therefore distant from this group of genes.)

$$a^+ \quad d^+$$

$$e^+ \quad a^+$$

$$c^+ \quad e^+$$

26. **a.** Owing to the medium used, all colonies are cys^+ but either + or − for the other two genes.

   b. (1) $cys^+\ leu^+\ thr^+$ and $cys^+\ leu^+\ thr^-$ (supplemented with threonine)

   (2) $cys^+\ leu^+\ thr^+$ and $cys^+\ leu^-\ thr^+$ (supplemented with leucine)

   (3) $cys^+\ leu^+\ thr^+$ (no supplements)

   c. Because none grew on minimal medium, no colony was $leu^+\ thr^+$. Therefore, medium (1) had $cys^+\ leu^+\ thr^-$, and medium (2) had $cys^+\ leu^-\ thr^+$. The remaining cultures were $cys^+\ leu^-\ thr^-$, and this genotype occurred in 100% − 56% − 5% = 39% of the colonies.

d. *cys* and *leu* are cotransduced 56% of the time, whereas *cys* and *thr* are cotransduced only 5% of the time. This indicates that *cys* is closer to *leu* than it is to *thr*. Since no *leu⁺ cys⁺ thr⁺* cotransductants are found, it indicates that *cys* is in the middle.

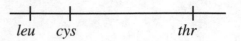

8

RECOMBINANT DNA AND GENETIC ENGINEERING

1. This problem assumes a random and equal distribution of nucleotides.

   *Alu*I $(1/4)^4$ = on average, once in every 256 nucleotide pairs

   *Eco*RI $(1/4)^6$ = on average, once in every 4096 nucleotide pairs

   *Acy*I $(1/4)^4(1/2)^2$ = on average, once in every 1024 nucleotide pairs

2. ***Unpacking the Problem***

   a. Of the two discussed in the text, pBR322 is the closer.

   b. The single *Hin*dIII site in pBP1 allows for a simple opening up of the plasmid so that a DNA fragment made with *Hin*dIII can be inserted.

   c. It is important because it allows for screening for insertions of DNA into the plasmid. If the plasmid simply recircularizes, the transformed bacteria will be tetracycline-resistant. If the plasmid contains "foreign" DNA, the *tet* gene will be disrupted and the strain will be tetracycline-sensitive.

   d. Insertion of donor DNA into the plasmid disrupts the *tet* gene. It is not relevant to the problem but was important in the construction of the library.

   e. A library is a large collection of cloned DNA maintained within easily cultured vectors. For this question, the source of donor DNA was *Hin*dIII-digested fruit fly genomic DNA, and the vector was pBP1. Although it is not relevant to this question, the source of donor DNA is often key to the type of research being conducted.

f.   The gene of interest would have been "found" by using a probe composed of the gene's sequence (typically just a small region is required). This could have been synthesized by "guessing" the DNA sequence on the basis of the gene product's amino acid sequence or by homology to a similar gene from another organism, etc. For this particular question, how this clone was identified does not matter.

g.   An electrophoretic gel is an apparatus to separate fragments of DNA by their size. Generally, the mix of DNA fragments is forced to migrate through an agarose gel by an electric field that is negative at the end where the DNA is placed and positive at the far end. Since DNA is negatively charged, it will move to the far end but at rates that are inversely proportional to its size: small fragments will move more rapidly than large.

h.   Ethidium bromide binds to DNA and fluoresces when exposed to UV light. It is used to visualize the location of the various DNA fragments within the gel.

i.   The DNA from this gel is not "blotted" onto filter paper in this problem. If it had been, it would have been a Southern blot (since DNA was in the gel).

j.   In this gel, DNA molecules of different sizes bound to ethidium bromide are visible under UV illumination.

k.   There is only one linear fragment generated when a circle is cut once.

l.   If cut twice, two linear fragments are generated.

m.   There is a one-to-one relationship between the number of sites cut in a circular plasmid and the number of fragments generated.

n.   Since the two enzymes will cut the DNA independently, the total number of fragments will be $n + m$.

o.   They were loaded into the wells located at the top of the diagram.

p.   Smaller fragments move more rapidly and travel farther per unit time than larger fragments.

q.   All the control lanes contain 5 kb of DNA, the size of the plasmid. Both *Hin*dIII and *Eco*RV cut the plasmid once but at separate locations, as seen in the lane of the double digest (both single digests generate a single band while the double digest generates two). The lanes with the clone 15-containing plasmid always add up to 7.5 kb, indicating that the donor DNA is 2.5 kb.

r.   No. The 5-kb plasmid is cut twice, and the resulting fragments must add up to the total length.

s.   No. They represent the cloned DNA cut out from the plasmid by *Hin*dIII and then cut once again by *Eco*RV. The sum of these fragments must equal the whole.

t.   It tells you that the fragment that disappeared also contains a restriction site for the second enzyme.

u. A probe will hybridize to any fragments to which it is complementary in sequence.

v. If the two vectors are nonhomologous, the only hybridization observed will be because the gene of interest from the one species is complementary to the gene of interest in the other.

Solution to the Problem

a.

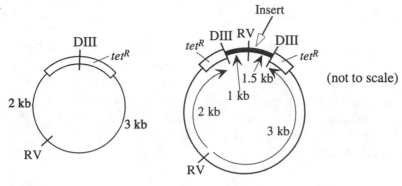

b. The tet^R gene used as a probe will detect only those bands that contain tet^R DNA. Thus, all bands in control lanes will have sequences complementary to the probe. For the clone 15 digests, the *Hind*III 5-kb band will be radioactive and ~~both~~ *NEITHER Eco*RV bands will be radioactive. For the *Hind*III + *Eco*RV double digest, the 3-kb and 2-kb bands will be radioactive.

c. The homologous gene used as a probe will detect only those fragments containing the gene of interest. Thus, no bands will be radioactive in the control lanes, and the clone 15 lanes will all have at least one radioactive band. For *Hind*III, the 2.5-kb band (the insert) will be radioactive. For *Eco*RV, the 4.5-kb and 3.0-kb bands will be radioactive. For the *Hind*III + *Eco*RV double digest, the 1.5-kb and 1-kb bands will be radioactive.

3. The data indicate that *Eco*RI fragments 1 and 4 contain no *Hind*II sites, fragment 3 contains one *Hind*II site, and fragment 2 contains two sites. Conversely, *Hind*II fragments A, B, and D all contain one *Eco*RI site, and fragment C contains none. Fragment D contains fragments 1 and 3_1; fragment A contains fragments 3_2 and 2_1; fragment C is the same as fragment 2_2; and fragment B contains fragments 2_3 and 4. The only map consistent with these data is

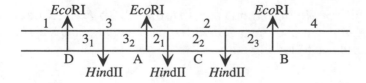

4. a. Since the actin protein sequence is known, a probe could be synthesized by "guessing" the DNA sequence based on the amino acid sequence. (This works best if there is a region of amino acids that can be coded with minimal redundancy.) Alternatively, the gene for actin cloned in another species can be used as a probe to find the homologous gene in *Drosophila*. If an expression vector was used, it might also be possible to detect a clone coding for actin by screening with actin antibodies.

b. Hybridization using the specific tRNA as a probe could identify a clone coding for itself.

5. To answer this problem, you must realize what is being visualized. The 8.5 *Eco*RI fragment is radioactive only at one 5′ end, and only fragments containing that end will be seen by autoradiography. When this fragment is cleaved by other restriction enzymes, the longest fragments will have been cut at sites farthest from the radioactive end. In the following figure, if cut at position labeled 2, the fragment will be longer than any fragment cut at 3, 4, or 5 and shorter than any cut at position 1.

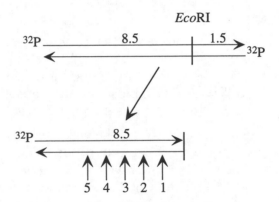

Using this logic, a relative map of the restriction sites of *Hin*dII and *Hae*III for this fragment can be generated.

Reading the gels from the top (and from the farthest to the nearest to the labeled end) the order is

(*Eco*RI) - *Hin*dII - *Hae*III - *Hae*III - *Hin*dII - *Hae*III - *Hae*III - *Hae*III - *Hin*dII - *Hin*dII - *Hae*III - *Hin*dII - *Hae*III - labeled end.

6. a. The double digest indicates that the 5.0-kb *Hin*dIII fragment also contains a *Sma*I site and the 5.5 *Sma*I fragment also contains a *Hin*dIII site. This suggests the following map:

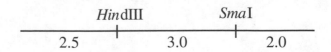

b. Since the only band to disappear is the 3.0-kb fragment, it is the only one that also contains an *Eco*RI site. The appearance of a new 1.5-kb fragment suggests the following:

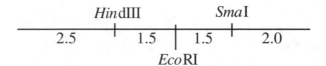

7. Create a library of *Podospora* DNA. This would be accomplished by isolating DNA from *Podospora*, cutting the DNA with *Hae*II, mixing with the vector pBR also cut with *Hae*II, ligating the mixture, and transforming *E. coli*. Only those bacteria that contain the plasmid will be *tet^R*. Of these, those that are *kan^S* contain plasmids with inserts.

     Assuming that the same genes from different species have approximately the same base sequence, the β-tubulin gene cloned from *Neurospora* can be used as a probe to isolate the β-tubulin gene from *Podospora*. Identify which clone or clones in the library contain the desired sequence by colony hybridization using the cloned *Neurospora* actin gene as a probe.

8. **a.** The transformed phenotype would map to the same locus. If gene replacement occurred by a double crossing-over event, the transformed cells would not contain vector DNA. If a single crossing-over took place, the entire vector would now be part of the linear *Neurospora* chromosome.

   b. The transformed phenotype would map to a different locus than that of the auxotroph if the transforming gene was inserted ectopically (i.e., at another location).

9. Size, translocations between known chromosomes, and hybridization to probes of known location can all be useful in identifying which band on a PFGE gel corresponds to a particular chromosome.

10. Conservatively, the amount of DNA necessary to encode this protein of 445 amino acids is 445 × 3 = 1335 base pairs. When compared with the actual amount of DNA used, 60 kb, the gene appears to be roughly 45 times larger than necessary. This "extra" DNA mostly represents the introns that must be correctly spliced out of the primary transcript during RNA processing for correct translation. (There are also comparatively very small amounts of both 5′ and 3′ untranslated regions of the final mRNA that are necessary for correct translation encoded by this 60-kb of DNA.)

11. The typical procedure is to "knock out" the gene in question and then see if there is any observable phenotype. One methodology to do this is described in Figure 8-30 in the companion text. A recombinant vector carrying a selectable gene within the gene of interest is used to transform yeast cells. Grown under appropriate conditions, yeast that have incorporated the marker gene will be selected. Many of these will have the gene of interest disrupted by the

selectable gene. The phenotype of these cells would then be assessed to determine gene function.

12.   a. There is one *Bgl*II site, and the plasmid is 14 kb.

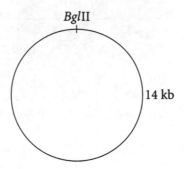

b. There are two *Eco*RV sites.

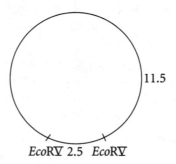

c. The 11.5 kb *Eco*RV fragment is cut by *Bgl*II. The arrangement of the sites must be as indicated below.

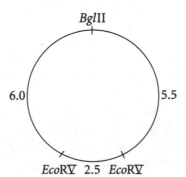

d. The *Bgl*II site must be within the *tet* gene.

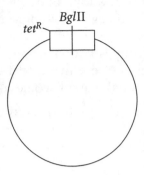

e. There was an insert of 4 kb.

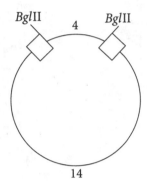

f. There was an *Eco*RV site within the insert.

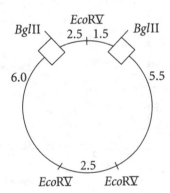

13. a. The restriction map of pBR322 with the mouse fragment inserted is shown below. The 2.5-kb and 3.5-kb fragments would hybridize to the pBR322 probe.

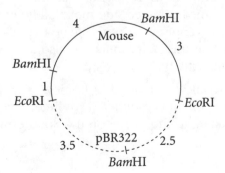

b. A protein 400 amino acids long requires a minimum of 1200 nucleotide bases. Only fragment 3 is long enough (3000 bp) to contain two or more copies of the gene. However, nothing can be said about their orientation.

14. a. To ensure that a colony is not, in fact, a prototrophic contaminant, the prototrophic line should be sensitive to a drug to which the recipient is resistant. A simple additional marker would also achieve the same end.

b. Use a nonrevertible auxotroph as the recipient (such as one containing a deletion).

15. a.

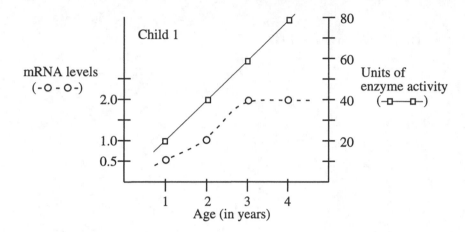

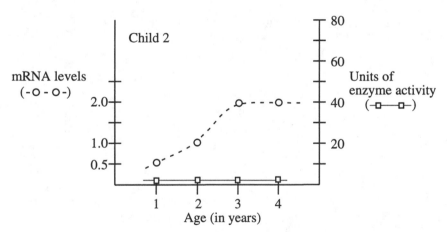

b., c. Very low levels of active enzyme are caused by the introduction of an *Xho* site within the D gene. Because the size of the nonfunctional enzyme has not changed, the most likely event was a point mutation within the coding region of the gene that also created a new *Xho* site. It is likely that this point mutation also altered the active site destroying enzyme activity.

d. Individual 1 would be defined as homozygous normal, while individual 2 is homozygous mutant. If either were heterozygous, there would be three bands hybridizing to the probe on the Southern blot.

16. a. The gel can be read from the bottom to the top in a 5´-to-3´ direction. The sequence is

$$5' \text{ TTCGAAAGGTGACCCCTGGACCTTTAGA } 3'$$

b. By complementarity, the template was

$$3' \text{ AAGCTTTCCACTGGGGACCTGGAAATCT } 5'$$

c. The double helix is

$$5' \text{ TTCGAAAGGTGACCCCTGGACCTTTAGA } 3'$$

$$3' \text{ AAGCTTTCCACTGGGGACCTGGAAATCT } 5'$$

d. Open reading frames have no stop codons. There are three frames for each strand, for a total of six possible reading frames. For the strand read from the gel, the transcript would be

$$5' \text{ UC}\underline{\textbf{UAA}}\text{AGGUCCAGGGGUCACCUUUCGAA } 3'$$

And for the template strand

$$5' \text{ UUCGAAAGG}\underline{\textbf{UGA}}\text{CCCCUGGACCUU}\underline{\textbf{UAG}}\text{A } 3'$$

Stop codons are in bold and underlined. There are a total of four open reading frames of the six possible.

17. The region of DNA that encodes tyrosinase in "normal" mouse genomic DNA contains two *Eco*RI sites. Thus, after *Eco*RI digestion, three different-sized fragments hybridize to the cDNA clone. When genomic DNA from certain albino mice is subjected to similar analysis, there are no DNA fragments that contain complementary sequences to the same cDNA. This indicates that these mice lack the ability to produce tyrosinase because the DNA that encodes the enzyme must be deleted.

18. Plant 1 shows the typical inheritance for a dominant gene that is heterozygous. Assuming kanamycin resistance is dominant to kanamycin sensitivity, the cross can be outlined as follows:

$$kan^R/kan^S \times kan^S/kan^S$$

$$\downarrow$$

$$^1/_2 \ kan^R/kan^S$$

$$^1/_2 \ kan^S/kan^S$$

This would suggest that the gene of interest would be inserted once into the genome.

Plant 2 shows a 3:1 ratio in the progeny of the backcross. This suggests that there have been two unlinked insertions of the kan^R gene and presumably the gene of interest as well.

$$kan^{R1}/kan^{S1} \; ; \; kan^{R2}/kan^{S2} \times kan^{S1}/kan^{S1} \; ; \; kan^{S2}/kan^{S2}$$

$$\downarrow$$

$$^1/_4 \quad kan^{R1}/kan^{S1} \; ; \; kan^{R2}/kan^{S2}$$

$$^1/_4 \quad kan^{R1}/kan^{S1} \; ; \; kan^{S2}/kan^{S2}$$

$$^1/_4 \quad kan^{S1}/kan^{S1} \; ; \; kan^{R2}/kan^{S2}$$

$$^1/_4 \quad kan^{S1}/kan^{S1} \; ; \; kan^{S2}/kan^{S2}$$

19. Assuming that the DNA from this region is cloned, it could be used as a probe to detect this RFLP on Southern blots. DNA from individuals within this pedigree would be isolated (typically from blood samples containing white blood cells) and restricted with *Eco*RI, and Southern blots would be performed. Individuals with this mutant CF allele would have one band that would be larger (owing to the missing *Eco*RI site) when compared with wild type. Individuals that inherited this larger *Eco*RI fragment would, at minimum, be carriers for cystic fibrosis. In the specific case discussed in this problem, a woman that is heterozygous for this specific allele marries a man that is heterozygous for a different mutated CF allele. Just knowing that both are heterozygous, it is possible to predict that there is a 25% chance of their child's having CF. However, since the mother's allele is detectable on a Southern blot, it would be possible to test whether the fetus inherited this allele. DNA from the fetus (through either CVS or amniocentesis) could be isolated and tested for this specific *Eco*RI fragment. If the fetus did not inherit this allele, there would be a 0% chance of its having CF. On the other hand, if the fetus inherited this allele, there would be a 50% chance the child will have CF.

20. The promoter and control regions of the plant gene of interest must be cloned and joined in the correct orientation with the glucuronidase gene. This places the reporter gene under the same transcriptional control as the gene of interest. Figure 8-33 in the companion text discusses the methodology used to create transgenic plants. Transform plant cells with the reporter gene construct, as discussed in the figure, grow into transgenic plants. The glucuronidase gene will now be expressed in the same developmental pattern as the gene of interest and its expression can easily be monitored by bathing the plant in an X-Gluc solution and assaying for the blue reaction product.

21. *Unpacking the Problem*

   a. Hyphae in *Neurospora* are the threads of cells that grow out from the original ascospore. Therefore, **hyphal extension** refers to the pattern of these threads and the distance that they grow. Since this process is due to cell growth (and its control) and cell shape, anyone interested in a vast array of cell biological issues might be interested in the genes identified by such screens.

b. **Mutational dissection** is the attempt to identify all the genes and gene products involved in a particular process. In this experiment, the goal is to identify all the genes that can mutate to a small-colony phenotype by random insertion of unrelated DNA (in this case a bacterial plasmid with a selectable marker).

c. *Neurospora* is a haploid organism, and this is relevant to this problem. What might otherwise be a recessive mutation in a diploid organism (typical of gene knockouts) would instead be immediately expressed in a haploid one.

d. The source of the DNA is not relevant to the problem as long as it contains a selectable marker for the organism being transformed. The ease of growing, manipulating, and isolating bacterial plasmids makes them an attractive choice.

e. Transformation, as originally discovered by Griffith, is the uptake of DNA from one organism by another organism and its ultimate expression. In this situation, *Neurospora* has been pretreated in such a way as to cause the uptake of the bacterial plasmid. This is a well-used technique in molecular genetics as a way of introducing genes from virtually any source into the organisms under study.

f. Plant and fungal cells are generally prepared for transformation by removal of their cell walls. The cell membranes are then exposed to a high salt concentration and the exogenous DNA. Studies indicate that the DNA enters the cell in two ways: (1) phagocytosis and (2) localized temporary dissolving of the membrane by the high salt concentration.

g. With successful transformation, the exogenous DNA passes through the cytoplasm and enters the nucleus, where it becomes integrated into a host chromosome.

h. Entry into Cell   Integration and Transformation

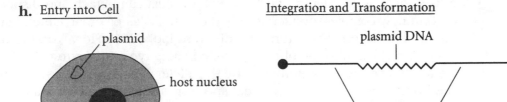

i. It is completely unnecessary to know what benomyl is. Its use simply allows for the selection of cells that received and integrated some exogenous DNA. Virtually any resistance marker could have been used. The choice of a resistance marker usually depends on what is easily available to the researcher, although questions of toxicity to humans may play a role in the choice.

j. Because hyphal extension occurs in colonies, not at the one-cell stage, the researcher must look for mutants that are expressed by a clone or colony. Therefore, he is looking for mutants that are "colonial." Mutations that produce an aberrant colony in size or shape are, by definition, involved with the extension of hyphae.

k. The "previous mutational analysis" could have been any random study. For example, in screens for specific auxotrophic mutants experimenters would have noticed this abnormal phenotype also appearing.

l.

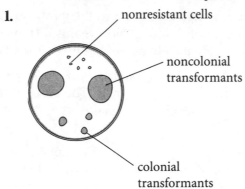

m. Tagging is a process to mutate and "mark" genes of interest by insertion of known DNA. In this case, the gene for benomyl resistance is being inserted randomly within the genome. Occasionally, insertion will occur within a gene involved with hyphal extension. These will cause the aberrant phenotype because the gene has been disrupted by the insertion of the selectable DNA. The disruption causes a knockout mutation within the gene of interest and also supplies a "molecular handle" to later clone the DNA.

n. The orange-colored bread mold *Neurospora* is a multicellular haploid in which the cells are joined end to end to form hyphae. The hyphae grow through the substrate and also send up aerial branches that bud off haploid cells known as conidia (asexual spores). Conidia can detach and disperse to form new colonies, or alternatively they can act as paternal gametes and fuse with a maternal structure of a different individual. However, the different individual must be of the opposite mating type. In *Neurospora* there are two mating types, determined by the alleles *A* and *a*. A cross will succeed only if it is *A* × *a*. An asexual spore from the opposite mating type fuses with a receptive hair, and a nucleus travels down the hair to pair with a maternal gamete that waits inside a specialized knot of hyphae. The *A* and *a* pair then undergo synchronous mitoses, finally fusing to form diploid meiocytes. Meiosis occurs, and in each meiocyte four haploid nuclei are produced, which represent the four products of meiosis. For an unknown reason, these four nuclei divide mitotically, resulting in eight nuclei, which develop into eight football-shaped sexual spores called ascospores. The ascospores are shot out of a flask-shaped fruiting body that has developed from the knot of hyphae that originally contained the maternal gametic cell. The ascospores can be isolated, each into a culture tube, where each ascospore will grow into a new culture by mitosis.

o. In order for the benomyl-resistance gene to be integrated within the host chromosome, it recombines with it. However, it is recombination between the *col* and *ben^R* genes that is interesting. It is either 0% (type 1), indicating that the insertion caused the small-colony phenotype, or 50% (type 2), showing that the two events are unlinked.

p. Only two types are possible: integration into a "hyphal" gene (so the resistance and small-colony phenotype are linked) and ectopic integration and concurrent mutation of a gene causing the small-colony phenotype where the two are unrelated and also unlinked. Which type is more likely depends on mutation rates and the number of genes that can mutate to the small-colony phenotype (i.e., the ease of generating spontaneous *col* mutants) and on the rate of transformation and integration.

q. A probe in experiments such as this one is usually a sequence of DNA that can be used to identify a specific DNA sequence within a genome or colony. The probe is labeled in some way to indicate its presence. In this experiment, the probe was probably the bacterial plasmid (although it might have been the benomyl-resistance gene), most likely radioactively labeled. A genomic library from each *col* mutant would be screened with the probe to identify those clones that contain complementary sequences (and, with luck, some sequences of the gene of interest).

r. A probe specific to the bacterial plasmid could be made by growing bacteria with the plasmid. The plasmids could be isolated through cesium chloride centrifugation and then labeled.

Solution to the Problem

a. Type 1 isolates behave genetically as if the benomyl-resistance and small-colony phenotype are completely linked. The progeny are all parental in phenotype (either *col ben-R* or *+ ben-S*). This would be the expected result if the insertion of the plasmid (and *ben-R* gene) caused the mutation that led to the *col* phenotype, which of course was the point of the experiment.

   Type 2 isolates behave genetically as if the benomyl-resistance and small-colony phenotype are unlinked (i.e., the two markers are segregating randomly in the progeny). This is the expected result if the insertion of the *ben-R* gene was unrelated to the mutation that led to the *col* phenotype. In other words, the *col* mutation occurred randomly and separately from the insertion of the *ben-R* gene.

b. The type 1 isolates are the *col* genes that are mutated by the insertion of the plasmid, and therefore these are the genes that are tagged.

c. The type 1 isolates should be used to create genomic libraries. The libraries should be screened for clones that contain DNA adjacent to the insert by probing with known sequences from the plasmid. The identified clones will represent parts of the disrupted gene of interest. To recover the intact wild-type gene, a subclone of the disrupted gene sequence can be used to probe a wild-type genomic library.

d. All progeny that are benomyl-resistant will also contain DNA from the bacterial plasmid integrated into their chromosome or chromosomes. Thus, all *ben-R* strains from this experiment will have DNA that hybridizes to a probe specific for the plasmid.

22. a., b. During Ti plasmid transformation, the kanamycin gene will insert randomly into the plant chromosomes. Colony A, when selfed, has $3/4$ kanamycin-resistant progeny, and colony B, when selfed, has $15/16$ kanamycin-resistant progeny. This suggests that there was a single insertion into one chromosome in colony A and two independent insertions on separate chromosomes in colony B. This can be schematically represented by showing a single insertion within one of the pair of chromosome "A" for colony A

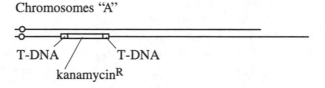

and two independent insertions into one of each of the pairs of chromosomes "B" and "C" for colony B.

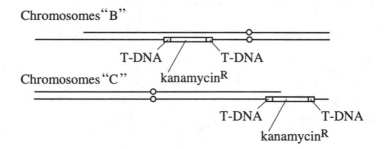

Genetically, this can be represented as

Colony A kan^{RA}/kan^{SA}

Colony B kan^{RB}/kan^{SB} ; kan^{RC}/kan^{SC}

When these are selfed $kan^{RA}/kan^{SA} \times kan^{RA}/kan^{SA}$

$$\downarrow$$

$1/4$ kan^{RA}/kan^{RA}

$1/2$ kan^{RA}/kan^{SA}

$1/4$ kan^{SA}/kan^{SA}

kan^{RB}/kan^{SB}; $kan^{RC}/kan^{SC} \times kan^{RC}/kan^{SB}$; kan^{RC}/kan^{SC}

↓

$9/16$ $kan^{RB}/-$; $kan^{RC}/-$

$3/16$ $kan^{RB}/-$; kan^{SC}/kan^{SC}

$3/16$ kan^{SB}/kan^{SB} ; $kan^{RC}/-$

$1/16$ kan^{SB}/kan^{SB} ; kan^{SC}/kan^{SC}

23. a. Yeast plasmids can exist free in the cytoplasm or can integrate into a chromosome; however, the patterns of inheritance will differ for the two states. Free plasmids are present in multiple copies and will be distributed to all progeny. This is observed in crosses with YP1 transformed cells: YP1 *leu⁺* × *leu⁻* all progeny are *leu⁺* and all have vector DNA. Crosses with YP2 transformed cells show simple Mendelian inheritance and suggest that this plasmid has integrated into the yeast chromosome.

b. For YP1 transformed cells, the circular (and free) plasmid will be linearized by the single restriction cut, and when probed, a single band will be present on the Southern blot. Since the YP2 plasmid is integrated, a single cut within the plasmid will generate two fragments that contain plasmid DNA. However, the size of these fragments will depend on where in the genome other sites exist for this same restriction enzyme. This is schematically shown below:

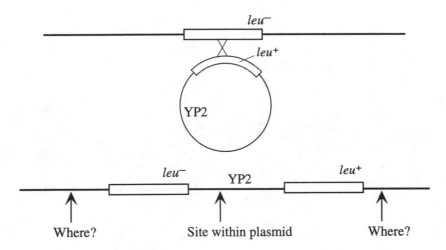

24. a. If the plasmid never integrates, the linear plasmid will be cut once by *Xba*I and two fragments will be generated that will both hybridize to the *Bgl*II probe. The autoradiogram will show two bands whose combined length will equal the full length of the plasmid.

b. If the plasmid integrates occasionally, most cells will still have free plasmids and these will be indicated by the two bands mentioned above. However, when the plasmid is integrated, two bands will still be generated, but their sizes will vary based on where other genomic *Xba*I sites are relative to the insertion point (see the figure for Problem 23b for similar logic). If integration is random, many other bands will be observed, but if it's at a specific site, only two other bands will be detected.

9
GENOMICS

1. The *Arabidopsis*-specific probe cross-hybridizes to DNA from cabbage. The number of bands observed is a function of where the specific restriction sites are relative to the region of DNA that hybridizes to the probe. Digestion with enzyme 2 results in a single band when hybridized to the probe because there are no restriction sites within the sequence that hybridizes. Digestion with enzyme 1 results in three bands because there are two restriction sites within the sequence that hybridizes. Similarly, digestion with enzyme 3 results in two bands when hybridized to the probe because there is one restriction site within this sequence. The schematic below shows an example.

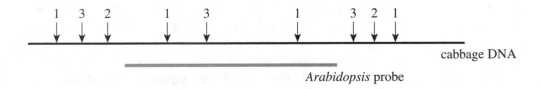

2. **a.** To determine the physical map showing the STS order, simply list the STSs that are positive, using parentheses if the order is unknown, and align them with one another to form a consistent order.

YAC A:			1	4	3	
YAC B:		5	1			
YAC C:				4	3	7
YAC D:	(6 2)	5				
YAC E:					3	7

b. Once the sequence of STSs is known, the YACs can be aligned as follows, although precise details of overlapping and the locations of ends are unknown:

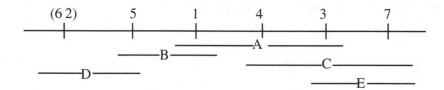

3. a. RAPDs are formed when regions of DNA are bracketed by two inverted copies of a "random" PCR primer sequence. Below, the primer is indicated by X's, and the amplified region appears in brackets. For convenience, the two amplified regions are shown on the same lengthy piece of DNA for strain 1.

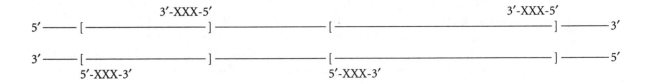

Strain 2 lacks one or two regions complementary to the primer.

b. Progeny 1 and 6 are identical with the strain 1 parent. Progeny 4 and 7 are identical with the strain 2 parent. Progeny 2 and 5 received the chromosome holding the upper band from the strain 1 parent and the chromosome holding the lower band from the strain 2 parent (resulting in no second band). Progeny 3 received the opposite: the chromosome holding the lower band from the strain 1 parent and the chromosome holding the upper band from the strain 2 parent (resulting in no second band). Therefore, bands 1 and 2 appear to be unlinked.

c. Recall that a nonparental ditype has two types only, both of which are recombinant. Therefore, the tetrad would be composed of two progeny like progeny 2 and two progeny like progeny 3.

4. a. The following stylized schematic of a reciprocal translocation between chromosome 3 and 21 is arbitrarily chosen to show the salient details. Band 3.1 of the q arm of chromosome 3 is split by the translocations that are correlated to the *N* disease allele. Probe c hybridizes to the region of 3q3.1 that remains with chromosome 3 and probes a, b, and d hybridize to the region of 3q3.1 that is translocated in this case to chromosome 21.

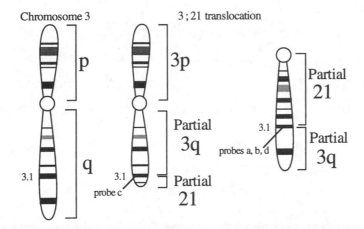

b. Since translocations of chromosome 3 that break band 3q3.1 are correlated to the disease, it is reasonable to assume that these rearrangements split the normal gene (n) in two, separating vital coding or regulatory regions. Therefore, analysis and cloning of this specific region should be attempted.

In order to isolate and characterize the normal allele, chromosome walking from the known clones should be attempted in genomic libraries from individuals with the translocation and affected with the disease. Probe c is to one side of the breakpoint, while a, b, and d are on the other side. Also, translocation breakpoints serve as useful molecular landmarks, since they are easily identified on Southerns as "split bands" when probed with cloned DNA spanning the breakpoint. Once the breakpoint has been identified and cloned, the appropriate subclones would be used to clone the normal allele from a "normal" genomic library. This would be in conjunction with the usual techniques to identify a gene: sequencing, open reading frame analysis, Northern blots, etc.

c. Once n is cloned, it can be used to clone the various alleles from individuals who have the disease but not a translocation. The various alleles could then be compared with n by sequence, regulation, etc.

5. From low to highest resolution the order would be:

f, c, (a, d), (e, h), b, g

6. For whole shotgun sequencing, the order would be:

e, b, h, g

7. This is just a matter of aligning the sequences to determine their overlap.

Read 1:   TGGCCGTGATGGGCAGTTCCGGTG

Read 2:   TTCCGGTGCCGGAAAGA

Read 3:   CTATCCGGGCGAACTTTTGGCCG

Read 4:   CGTGATGGGCAGTTCCGGTG

Read 5:   TTGGCCGTGATGGGCAGTT

Read 6:   CGAACTTTTGGCCGTGATGGGCAGTTCC

And this creates the contig:

CTATCCGGGCGAACTTTTGGCCGTGATGGGCAGTTCCGGTGCCGGAAAGA

8.

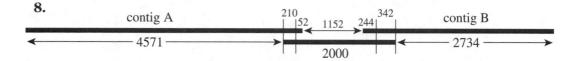

9. You can determine whether the cDNA clone was a monster or not by alignment of the cDNA sequence against the genomic sequence. (There are computer programs available to do this.) Is it derived from two different sites? Does the cDNA map within one (gene-sized) region in the genome or to two different regions? Of course, introns may complicate the issue.

10. a. Since the triplet code is redundant, changes in the DNA nucleotide sequence (especially at those nucleotides coding for the third position of a codon) can occur without change to its encoded protein.

  b. It can be expected that protein sequences will evolve and diverge more slowly than the genes that encode them.

11. a., b. There are four patterns that can be observed in the comparisons that can be made between these six markers: + +, − −, + −, and − +. The first two indicate concordance and the second two indicate a lack of concordance. Ideally, data would show either 100% concordance for the seven hybrids, indicating linkage, or 100% discordance for the seven hybrids, indicating a lack of linkage.

Because radiation hybrids involve chromosome breakage, two genes that are located very close together on the same chromosome may show some discordance despite the close linkage. Two genes that are located on different chromosomes may also show some concordance due to the chance that two separate fragments may become established within a single hybrid line. Therefore, the problem is how to distinguish between reduced concordance due to chromosome fragmentation and chance concordance due to two fragments from different chromosomes being in the same hybrid. Obviously, a statistical solution is needed, but there are not enough data in this problem for a statistical analysis.

Sort the data into three groups: 100% concordance, 100% discordance, and mixed (concordance/discordance). This follows below:

100% Concordance	100% Discordance	Mixed
E-F	None	A-B 2/5
		A-C 2/5
		A-D 6/1
		A-E 2/5
		A-F 2/5
		B-C 5/2
		B-D 1/6
		B-E 3/4
		B-F 3/4
		C-D 3/4
		C-E 3/4
		C-F 3/4
		D-E 3/4
		D-F 3/4

Markers E and F are most likely located on the same chromosome. Markers B and D may be located on different chromosomes.

In the absence of statistical analysis, with so few total hybrids, it is important to pay more attention to the + + patterns than the − − patterns simply because − − can arise either from linkage, with the specific chromosome missing in the hybrid, or from lack of linkage, with the two chromosomes (or fragments) lacking in the hybrid. Therefore, going back to the mixed category and focusing on those marker pairs that had a high degree, but not 100%, of concordance, one sees that the 6/1 pattern of A-D and the 5/2 pattern of B-C stand out. For the A-D pair, 3 of the 6 concordances are + +, while only 2 of the 5 concordances for B-C are + +. It is unclear from the data whether this is a significant difference, and significance cannot be determined in any fashion. Therefore, it would be important to collect more data before drawing further conclusions.

12. *Unpacking the Problem*

a. Two types of hybridizations that have already been discussed are hybridizations between strains of a species and hybridizations between species. A third type of hybridization is referred to in this problem: molecular hybridization. Molecular hybridization can involve either DNA-DNA hybridization or

DNA-RNA hybridization. In both instances, it relies on the specificity of complementary pairing and can take place in solution, on a gel, on a filter, or on a slide. For example:

5′—UACGGGAU—3′ RNA

3′—ATGCCCTA—5′ DNA

b. In situ hybridization usually is conducted on a slide so that the stained chromosomes can be observed and the specific portion of a chromosome to which the probe hybridizes can be identified.

c. A YAC is a yeast artificial chromosome. It contains a yeast centromere, autonomous replication sequences (origins of replication), telomeres, and DNA that has been attached between them.

d. Chromosome bands are dark regions along the length of a chromosome that occur in a characteristic pattern for each chromosome within an organism. They can occur naturally, as with *Drosophila* polytene chromosomes, or they can be induced by a number of chemical and physical agents, combined with staining to accentuate the bands and interbands.

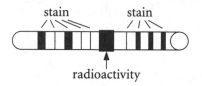

e. The five YACs could have been hybridized sequentially to the same chromosome preparation, which is, however, unlikely. Alternatively, the information could have been determined in five separate experiments. In either case, a YAC labeled with either radioactivity or fluorescence, and including the DNA of interest, was hybridized to a chromosome preparation. The chromosomes were properly treated to reveal the banding pattern, and the YACs were determined to hybridize to the same band.

f. A genomic fragment, by definition, contains a subportion of the genome being studied. In most instances, it actually contains a subportion of one chromosome. Five randomly chosen YACs would not be expected to contain the same genomic fragment or even fragments from the same chromosome. The fragments could have been produced by either physical (X-irradiation, shearing) or chemical (digestion, restriction) means, but it does not matter how they were produced.

g. A restriction enzyme is a naturally occurring bacterial enzyme that is capable of causing either single- or double-stranded breaks in DNA at specific DNA sequences.

h. A long cutter is a restriction enzyme that produces very long fragments of DNA because the sequence it recognizes occurs infrequently within the genome.

i. The YACs were radioactively labeled so that their location after hybridization could be detected through autoradiography. To radioactively label is to attach an isotope that emits energy through decay. Commonly used radioactive labels are tritium (^{3}H) in place of hydrogen and ^{32}P in place of phosphorus.

j. An autoradiogram is a "self-picture" taken through radioactive decay from a labeled probe. When a gel or blot is used, the radioactive decay is captured by a piece of X-ray film. When in situ hybridization is performed on slides, the photographic emulsion coats the slide directly.

k. Free choice. Be sure you truly know the meaning of each term.

l. We are given a diagram of the composite autoradiographic results. The DNA from humans was isolated and subjected to digestion by a restriction enzyme that cuts very infrequently. Once the DNA was electrophoresed, it was Southern-blotted and then probed sequentially with radioactively labeled YACs, followed by sequential exposure to X-ray film. Between probings, the previous YAC hybrid was removed through denaturing of the DNA-DNA hybrid. Alternatively, five separate Southern blottings were done.

m. The haploid human genome is thought to contain approximately 3.3×10^6 kilobases of DNA .

n. Restriction digestion of human genomic DNA would be expected to produce hundreds of thousands of fragments.

o. The fragments produced by restriction of human genomic DNA would be expected to be mostly different.

p. When subjected to electrophoresis and then stained with a DNA stain, the digested human genome would produce a continuous "smear" of DNA, from very large fragments (in excess of tens of thousands of base pairs in length) to fragments that are very small (under a hundred bases in length).

q. In this question, only two distinct bands are produced, at most, in any one probing. The difference between what is seen with a DNA stain and what is seen with probing lies in the specificity of the agent being used. DNA stain will detect any DNA, whereas a DNA probe will detect only DNA that is complementary to the probe.

r. Number them from top to bottom, 1–3, across the gel. Thus, YACs A–C contain band 1, YACs C–D contain band 2, and YACs A and E contain band 3.

s. There are no restriction fragments on the autoradiogram. The fragments are on the filter (nitrocellulose, nylon) used to blot the gel. The radioactivity of the probes is captured by the X-ray film as it decays, producing an exposed region of film.

t. YACs B, D, and E hybridize to one fragment, and YACs A and C hybridize to two fragments.

u. A YAC can hybridize to two fragments if the YAC contains continuous DNA and there is a restriction site within that region. A YAC can also hybridize to two fragments if it contains discontinuous DNA from two locations in the genome that either are on different chromosomes (this is analogous to a translocation) or are separated by at least two restriction sites if they are on the same chromosome (this is analogous to a deletion). In this case, the former makes more sense. Because the YACs were selected for their binding to one specific chromosome band, it is unlikely that the YACs are composed of discontinuous DNA sequences. A YAC could hybridize to more than two fragments because the continuous DNA could contain many restriction sites or the discontinuous DNA could be composed of DNA from a number of regions in the genome.

v. Cytogeneticists use the term *band* to designate a region of a chromosome that is dark-staining. Molecular biologists use *band* to designate a region of dark appearing on an autoradiogram, which is produced by radioactive decay from a specific probe that reacted with a population of molecules localized by gel electrophoresis. In both cases, *band* refers to a localization.

Solution to the Problem

a. Note that fragments 1 and 3 occur together and fragments 1 and 2 occur together, but that fragments 2 and 3 do not occur together. This suggests that the sequence is 2 1 3 (or 3 1 2).

b. If the sequence of the fragments is 2 1 3, then the YACs can be shown in relation to these fragments. YAC A spans at least a portion of both 1 and 3. YAC B is within region 1. YAC C spans at least a portion of regions 1 and 2. YAC D is contained within region 2. YAC E is contained within region 3. A diagram of these results is shown below. In the diagram, there is no way to know the exact location of the ends of each YAC.

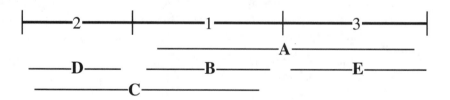

13. a. This is just a matter of aligning the sequences to determine their overlap.

Read 1:   ATGCGATCTGTGAGCCGAGTCTTTA

Read 2:   AACAAAAATGTTGTTATTTTTATTTCAGATG

Read 3:   TTCAGATGCGATCTGTGAGCCGAG

Read 4:   TGTCTGCCATTCTTAAAAACAAAAATGT

Read 5:   TGTTATTTTTATTTCAGATGCGA

Read 6:   AACAAAAATGTTGTTATT

And this creates the contig:

TGTCTGCCATTCTTAAAAACAAAAATGTTGTTATTTTTATTTCAGATGCGATCTGTGAGCCGAGTCTTTA

and the transcript of:

UGUCUGCCAUUCUUAAAAACAAAAAUGUUGUUAUUUUUAUUUCAGAUGCGAUCUGUGAGCCGAGUCUUUA

b. Translation of the contig starting at the first letter would give:

      CLPFLKTKMLLFLFQMRSVSRVF

Translation of the contig starting at the second letter would give:

      VCHS stop

Translation of the contig starting at the third letter would give:

      SAILKNKNVVIFISDAICRPSL

c. Using the nucleotide sequence of the contig and performing BLASTn or the possible translation products and performing tBLASTn, you will discover that this sequence and the translation product above listed first match perfectly with a region of exon 19 of the human CFTR gene.

14. The cross is

$$cys\text{-}1 \; RFLP\text{-}1^O \; RFLP\text{-}2^O \quad \times \quad cys\text{-}1^+ \; RFLP\text{-}1^M \; RFLP\text{-}2^M$$

Scoring the progeny, a parental type will have the genotype of either strain and, if the markers are all linked, be the most common. A recombinant type will have a mixed genotype and be less common. Clearly, the first two ascospore types are parental, with the remaining being recombinant.

a. The *cys-1* locus is in this region of chromosome 5. If it were not in this region, linkage to either of the RFLP loci would not be observed.

b. To calculate specific distances, you may need to review previous chapters. Here, it is assumed that you recall basic mapping strategies.

cys-1 to *RFLP-1* = $^{(2 + 3)}/_{100}$ × 100% = 5 map units

cys-1 to *RFLP-2* = $^{(7 + 5)}/_{100}$ × 100% = 12 map units

RFLP-1 to *RFLP-2* = $^{(2 + 3 + 7 + 5)}/_{100}$ × 100% = 17 map units

```
      |—5 m.u.—|————12 m.u.————|
    RFLP-1   cys-1           RFLP-2
```

c. A number of strategies could be tried. Since this is an auxotrophic mutant, functional complementation can be attempted. Positional cloning or chromosome walking from the RFLPs is also a very common strategy.

15. The correct assembly of large and nearly identical regions is problematic with either method of genomic sequencing. However, the whole genome shotgun method is less effective at finding these regions than the clone-based strategy.

This method also has the added advantage of easy access to the suspect clone(s) for further analysis.

16. a. Of the regions of overlap for cosmids C, D, and E, region 5 is the only region in common. Thus, gene x is localized to region 5.

   b. The common region of cosmids E and F, or the location of gene y, is region 8.

   c. Both probes are able to hybridize with cosmid E because the cosmid is long enough to contain part of genes x and y.

17. a. DNA from each individual was obtained. It was restricted, electrophoresed, blotted, and then probed with the five probes. After each probing, an autoradiograph was produced.

   b. First, identify which chromosome came from the affected parent. This is easily determined by identifying which chromosome could not have come from the mother. For the first daughter, the chromosome with 2´ was inherited from the father. Likewise, 2´´, 3´´, and 2´´ identify the paternal chromosome in the other children. In all cases, the chromosome drawn to the left is the one inherited from the mother.

   Next compare the maternal chromosomes of affected offspring with unaffected offspring to determine which RFLP is most closely correlated to the disease. This analysis is based on the co-segregation of one of the RFLPs and the disease-causing gene. Notice that all of these chromosomes show evidence of recombination. For example, when compared with the mother's chromosomes, it can be deduced that the maternally inherited chromosome of the unaffected daughter is the result of a double crossover event.

Affected:	1° 2° 3´ 4° 5°
Unaffected:	1° 2° 3´ 4´ 5°
Unaffected:	1´ 2° 3´ 4´ 5°
Affected:	1´ 2° 3´ 4° 5´

   The only RFLP that correlates to the disease and therefore is likely closest to the disease allele is 4°. It is present in both affected children and absent in both unaffected children.

   c. It appears that RFLP 4 is the closest marker to the gene and could be used for positional cloning by chromosome walking. However, with only four offspring, the genetic distance between the gene and this marker could be quite large. The number of markers for each human chromosome is already large and increasing almost daily. If possible, it makes sense to further analyze this family (and as many other families with the same trait that can be found) to see if the gene can be further localized before the arduous task of "walking" is attempted.

18. **a.** **Clone fingerprinting:** clones are digested with restriction enzyme(s) to generate a set of bands whose number and positions are a unique "fingerprint" of that clone; the different bands generated from separate clones can be aligned to determine if there is any overlap between the inserted DNAs; the overlap is used to generate contigs.

    STS content mapping: amass a large set of random clones with small genomic inserts; sequence short regions of each clone and design pairs of PCR primers based on these sequences to amplify the short DNA sequence flanked by the primers (these short sequences are called STSs or sequence-tagged sites); characterize clones of large genomic inserts to identify the contained STSs. Clones that are shown to have specific STSs in common must be overlapping and can therefore be aligned into contigs.

    Radiation hybrid mapping: irradiate cells to fragment chromosomes; fuse irradiated cells with rodent cells to form a panel of different hybrids (each hybrid will have a random assortment of fragments integrated into the rodent chromosomes); analyze different radiation hybrids for co-occurrence of markers, which may indicate linkage.

b. If two different clones have repetitive transposable element sequences in common, they will also share restriction fragments (bands) from those elements. If the randomly chosen STS is from a repeated element, its presence in various clones will not necessarily indicate the clones overlap. Co-occurrence of markers that by chance are repetitive will not necessarily indicate linkage in radiation hybrid mapping.

c. Attempt to use restriction enzymes that do not cut within repeated elements. Make certain that the STSs are unique (present just once in the genome). Make certain that markers being assessed for co-occurrence are not repeated.

19. **a., b., c.** Cystic fibrosis (CF) is a recessive, autosomally inherited disease. Both parents in this pedigree must be carriers since some of their children are affected. Because the problem states that the three probes used are very closely linked to the *CF* gene, recombination will be ignored.

    The data from three probes are presented, but only probes 1 and 3 detect RFLPs in this pedigree and are therefore informative. Both probes detect either one or two bands depending on the allele present. Calling the one-band pattern allele *A* and the two-band pattern allele *B*, the individuals of the pedigree are

Father	*RFLP-1^B RFLP-3^A*
Mother	*RFLP-1^A RFLP-3^B*
Child 1 (II-1)	*RFLP-1^B RFLP-3^A* (does not have CF)
Child 2 (II-2)	*RFLP-1^B RFLP-3^B* (does have CF)
Child 3 (II-3)	*RFLP-1^B RFLP-3^B* (does have CF)
Child 4 (II-4)	*RFLP-1^A RFLP-3^B* (does not have CF)
Child 5 (II-5)	*RFLP-1^B RFLP-3^A* (does not have CF)
Child 6 (II-6)	*RFLP-1^B RFLP-3^B* (does have CF)
Child 7 (II-7)	*RFLP-1^A RFLP-3^A* (does not have CF)

The first step is to determine which RFLP alleles are linked to the disease-causing *CF* alleles. The pattern of inheritance suggests that *RFLP-1^B* from the father and *RFLP-3^B* from the mother are both linked to *CF* alleles since all children that are *RFLP-1^B RFLP-3^B* also have CF. The oldest son (II-1) is a carrier since he has inherited a *CF* allele (linked to *RFLP-1^B*) from his father. Similarly, II-4 has inherited a *CF* allele from his mother, II-5 has inherited a *CF* allele from his father, and II-7 is homozygous normal.

20. Assessing whether a short sequence constitutes an exon is difficult. Identification of consensus donor and acceptor splice site sequences can be tried and also the use of comparative genomics, that is, the conservation of the predicted amino acid encoded by the micro-exon in the same or other genomes.

10 GENE MUTATIONS:
ORIGINS AND REPAIR PROCESSES

1. **a.** A transition mutation is the substitution of a purine for a purine or the substitution of a pyrimidine for a pyrimidine. A transversion mutation is the substitution of a purine for a pyrimidine, or vice versa.

   b. Both are base-pair substitutions. A synonymous mutation is one that does not alter the amino acid sequence of the protein product from the gene, because the new codon codes for the same amino acid as did the nonmutant codon. A neutral mutation results in a different amino acid that is functionally equivalent, and the mutation therefore has no known adaptive significance.

   c. A missense mutation results in a different amino acid in the protein product of the gene. A nonsense mutation causes premature termination of translation, resulting in a shortened protein.

   d. Frameshift mutations arise from addition or deletion of one or more bases in other than multiples of three, thus altering the reading frame for translation. Therefore, the amino acid sequence from the site of the mutation to the end of the protein product of the gene will be altered. Frameshift mutations can and often do result in premature stop codons in the new reading frame, leading to shortened protein products. A nonsense mutation causes premature termination of translation, resulting in a shortened protein.

2. Frameshift mutations arise from addition or deletion of one or more bases in other than multiples of three. When translated, this will alter the reading frame and therefore the amino acid sequence from the site of the mutation to the end of the protein product. Also, frameshift mutations often result in premature stop codons in the new reading frame, leading to shortened protein products. A missense mutation changes only a single amino acid in the protein product.

3. Misalignment of homologous chromosomes during recombination results in a duplication in one strand and a corresponding deletion in the other.

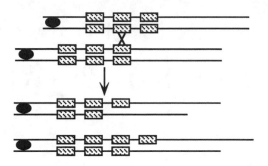

Recombination between two homologous repeats in a looped DNA molecule can lead to deletion.

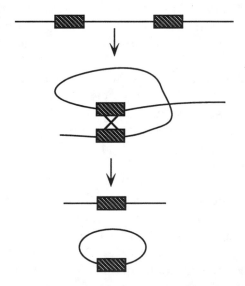

Also, "slipped mispairing" (see Figure 10-12 in the companion text) may result in a deletion in the newly synthesized DNA strand during replication.
All these mechanisms are supported by DNA sequencing results.

4. Depurination results in the loss of the adenine or guanine base from the DNA backbone. Since the resulting apurinic site cannot specify a complementary base, replication is blocked. Under certain conditions, replication proceeds with

a near random insertion of a base opposite the apurinic site. In three-fourths of these insertions, a mutation will result.

Deamination of cytosine yields uracil. If left unrepaired, the uracil will be paired with adenine during replication, ultimately resulting in a transition mutation. Deamination of 5-methylcytosine yields thymine and thus frequently leads to C → T transitions.

Oxidatively damaged bases, such as 8-oxodG (8-oxo-7-hydrodeoxyguanosine) can pair with adenine, resulting in a transversion.

Errors during DNA replication (see Figure 10-12 in the companion text) can lead to spontaneous indel mutations.

5. 5-BU is an analog of thymine. It undergoes tautomeric shifts at a higher frequency than does thymine and therefore is more likely to pair with G than thymine is during replication. At the next replication this will lead to a GC pair rather than the original AT pair. On the other hand, 5-BU can also be incorporated into DNA by mispairing with guanine. In this case, it will convert a GC pair to an AT pair.

EMS is an alkylating agent that produces O-6-ethylguanine. This alkylated guanine will mispair with thymine, which leads from a GC pair to an AT pair at the next replication.

6. An AP site is an apurinic or apyrimidinic site. AP endonucleases introduce chain breaks by cleaving the phosphodiester bonds at the AP sites. Some exonuclease activity follows, so that a number of bases are removed. The resulting gap is filled by DNA pol I and then sealed by DNA ligase.

General excision repair is used to remove damaged DNA including photodimers. It cleaves the phosphodiester backbone on either side of the damage, removing 12 or 13 nucleotides in bacteria and 27 to 29 nucleotides in eukaryotes. The resulting gap is filled by DNA pol I and then sealed by DNA ligase.

Photodimers can be also be repaired by photolyase (found in bacteria and lower eukaryotes), which binds to and splits the dimer in the presence of certain wavelengths of light. In bacteria the damaged DNA can also be bypassed by the SOS system, which enables DNA polymerase to "fill in" random bases where it encounters photodimers on the template strand. Finally, although photodimers on the template strand cause DNA pol III to stall, replication can restart downstream from the dimer, leaving a region of single-stranded DNA. The single-stranded DNA will attract single-stranded-binding protein and another protein, RecA, which can effect recombinational repair using DNA from the sister chromatid to patch the gap.

7. Mismatch repair occurs if a mismatched nucleotide is inserted during replication. The new, incorrect base is removed and the proper base is inserted. The enzymes involved can distinguish between new and old strands because, in *E. coli,* the old strand is methylated.

Recombination repair occurs if lesions such as AP sites and UV photodimers block replication (there is a gap in the new complementary strand). Recombination fills this gap with the corresponding segment from the sister DNA molecule, which is normal in both strands. This produces one DNA molecule with a gap across from a correct strand, which can then be filled by complementation, and one with a photodimer across from a correct strand.

8. Yes. It will cause CG → TA transitions.

9. Depurination results in the formation of an AP site. AP endonucleases introduce chain breaks by cleaving the phosphodiester bonds at these sites. Some exonuclease activity follows, so that a number of bases are removed. The resulting gap is filled by DNA pol I and then sealed by DNA ligase.

Deamination of cytosine and adenine yields uracil and hypoxanthine, respectively. Specific glycosylases remove these bases, creating an AP site which is then repaired as above. Deamination of 5-methylcytosine yields thymine, and since this cannot be distingushed from any other thymine in the DNA, 5-methylcytosines represent mutational "hot spots."

10. In replicative transposition, transposable elements move to a new location by replicating into the target DNA, leaving behind a copy of the transposable element at the original site. If, on the other hand, the transposable element excises from its original position and inserts into a new position, this is called conservative transposition.

11. The Streisinger model proposed that frameshifts arise when loops in single-stranded regions are stabilized by slipped mispairing of repeated sequences. In the *lac* gene of *E. coli,* a four-base-pair sequence is repeated three times in tandem, and this is the site of a hot spot.

The sequence is 5′-CTGG CTGG CTGG-3′. During replication the DNA must become single stranded in short stretches for replication to occur. As the new strand is synthesized, transient disruptions of the hydrogen bonds holding the new and old strands together may be stabilized by the incorrect base pairing of bases that are now out of register by the length of the repeat, or in this case, a total of four bases. Depending on which strand, new or template, loops out with respect to the other, there will be an addition or deletion of four bases, as diagrammed below:

```
            T G

         C       G

   5′-C T G G        C T G G-3′      →      DNA synthesis

   3′-G A C C ——————— G A C C  G A C C-5′
```

In this diagram, the upper strand looped out as replication was occurring. The loop is stabilized by base pairing on either strand. As replication continues at the 3′ end, an additional copy of CTGG will be synthesized, leading to an addition of four bases. This will result in a frameshift mutation.

12. **a.** Because 5´-UAA-3´ does not contain G or C, a transition to a GC pair in the DNA cannot result in 5´-UAA-3´. 5´-UGA-3´ and 5´-UAG-3´ have the DNA antisense-strand sequence of 3´-ACT-5´ and 3´-ATC-5´, respectively. A transition to either of these stop codons occurs from the nonmutant 3´-ATT-5´, respectively. However, a DNA sequence of 3´-ATT-5´ results in an RNA sequence of 5´-UAA-3´, itself a stop codon.

   b. Yes. An example is 5´-UGG-3´, which codes for Trp, to 5´-UAG-3´.

   c. No. In the three stop codons the only base that can be acted upon is G (in UAG, for instance). Replacing the G with an A would result in 5´-UAA-3´, a stop codon.

13. Mutations in gal can be generated and from these strains λ*dgal* phage isolated. Through hybridization of denatured λ*dgal* DNA containing the mutation with wild-type λ*dgal* DNA, some of the molecules will be heteroduplexes between one mutant and one wild-type strand. If the mutation was caused by an insertion, the heteroduplexes will show a "looped out" section of single-stranded DNA, confirming that one DNA strand contains a sequence of DNA not present in the other (see Figure 10-24 in the companion text).

   Page 331 of your text describes a method to compare the densities of *gal*⁺-carrying λ phage with *gal*⁻-carrying phage. In this experiment, the *gal*⁻-phage are denser, indicating that they contain a larger DNA molecule.

   If the *gal* genes are cloned, direct comparison of the restriction maps or even the DNA sequence of mutants compared with wild type will give specific information about whether any are the results of insertions.

14. **a., b.** Mutant 1: most likely a deletion. It could be caused by radiation.

   Mutant 2: because proflavin causes either additions or deletions of bases and because spontaneous mutation can result in additions or deletions, the most probable cause was a frameshift mutation by an intercalating agent.

   Mutant 3: 5-BU causes transitions, which means that the original mutation was most likely a transition. Because HA causes GC → AT transitions and HA cannot revert it, the original must have been a GC → AT transition. It could have been caused by base analogs.

   Mutant 4: the chemical agents cause transitions or frameshift mutations. Because there is spontaneous reversion only, the original mutation must have been a transversion. X-irradiation or oxidizing agents could have caused the original mutation.

   Mutant 5: HA causes transitions from GC → AT, as does 5-BU. The original mutation was most likely an AT → GC transition, which could be caused by base analogs.

   c. The suggestion is a second-site reversion linked to the original mutant by 20 map units and therefore most likely in a second gene. Note that auxotrophs equal half the recombinants.

15. **a.** A lack of revertants suggests either a deletion or an inversion within the gene.

b. To understand these data, recall that half the progeny should come from the wild-type parent.

Prototroph A: because 100% of the progeny are prototrophic, a reversion at the original mutant site may have occurred.

Prototroph B: half the progeny are parental prototrophs, and the remaining prototrophs, 28%, are the result of the new mutation. Notice that 28% is approximately equal to the 22% auxotrophs. The suggestion is that an unlinked suppressor mutation occurred, yielding independent assortment with the *nic* mutant.

Prototroph C: there are 496 "revertant" prototrophs (the other 500 are parental prototrophs) and 4 auxotrophs. This suggests that a suppressor mutation occurred in a site very close [$100\%(4 \times 2)/1000 = 0.8$ m.u.] to the original mutation.

16. O^{-6}-methyl G leads predominantly to high levels of GC $\rightarrow$ AT transitions. 8-oxodG gives rise predominantly to high levels of G $\rightarrow$ T transversions. Finally, C-C photodimers will most often cause C $\rightarrow$ T transitions, but some transversions are also possible.

17. Compare the original amino acid sequences to the mutant ones and list the changes.

> original: ile; mutant: met
>
> original: asn; mutant: ser
>
> original: stop; mutant: trp

Now compare the codons that must have been altered by this mutagen.

> original: ile AU<u>A</u>; mutant: met AU<u>G</u>;
>
> original: asn A<u>A</u>C or A<u>A</u>U; mutant: ser A<u>G</u>C or A<u>G</u>U
>
> original: stop U<u>A</u>G or UG<u>A</u>; mutant: trp U<u>G</u>G;

All these mutations can be the result of a T $\rightarrow$ C transition in the DNA. This would result in the A $\rightarrow$ G change in the mRNA that explains all three codon changes. This mutagen, then, might work by altering the base-pairing specificity of T so that it now base pairs with G.

18. Yes. Because DNA is a double-stranded molecule, replication of the DNA strand with a T $\rightarrow$ T* (altered T that base pairs with G) change produces an A to G transition in the newly replicated complementary DNA strand. If the mRNA is transcribed from the strand with the A $\rightarrow$ G change (the template strand), a U $\rightarrow$ C change is produced in the corresponding mRNA.

> original: leu C<u>U</u>N; mutant: pro C<u>C</u>N (where N = any base)

19. Plasmids that contain drug-resistance genes are called R factors and, like F factors, these self-replicating plasmids are transferred rapidly by cell conjugation. Drug-resistance genes reside between two identical IS-like elements called IR sequences. Collectively, the IR sequences with their contained genes are called a transposon. Transposons are mobile elements and as such can "jump" from a plasmid to a bacterial chromosome or from one plasmid to another plasmid. This combination of drug-resistance gene mobility and R plasmid cell-to-cell transfer quickly leads to the generation of multiple-drug-resistance plasmids given the "right" environment...like a hospital!

20. P elements are transposable elements found in *Drosophila*. They consist of a gene coding for transposase (the enzyme required for transposition) flanked by small inverted repeats. Many P elements found in a typical genome have deletions removing some or all of the coding information for the transposase and therefore are not capable of transposition unless the enzyme is supplied from another element or from a transgene. The movement of these nonautonomous elements, then, can be controlled by doing crosses in which a source of the transposase is introduced into a genome containing only deleted elements. Their controlled transposition creates mutations by insertion, and these interrupted and mutated genes can then be cloned with the use of P element segments as a probe—a method called "tagging."

11 CHROMOSOME MUTATIONS

1. **a.** Cytologically, deletions lead to shorter chromosomes with missing bands (if banded) and an unpaired loop during meiotic pairing when heterozygous. Genetically, deletions are usually lethal when homozygous, do not revert, and when heterozygous, lower recombinational frequencies and can result in "pseudodominance" (the expression of recessive alleles on one homolog that are deleted on the other). Occasionally, heterozygous deletions express an abnormal (mutant) phenotype.

   b. Cytologically, duplications lead to longer chromosomes and, depending on the type, unique pairing structures during meiosis when heterozygous. These may be simple unpaired loops or more complicated twisted loop structures. Genetically, duplications can lead to asymmetric pairing and unequal crossing-over events during meiosis, and duplications of some regions can produce specific mutant phenotypes.

   c. Cytologically, inversions can be detected by banding, and when heterozygous, they show the typical twisted "inversion" loop during homologous pairing. Pericentric inversions can result in a change in the p:q ratio. Genetically, no viable crossover products are seen from recombination within the inversion when heterozygous, and as a result, flanking genes show a decrease in RF.

   d. Cytologically, reciprocal translocations may be detected by banding, or they may drastically change the size of the involved chromosomes as well as the positions of their centromeres. Genetically, they establish new linkage relationships. When heterozygous, they show the typical cross structure during meiotic pairing and cause a diagnostic 50% reduction of viable gamete production, leading to semisterility.

2. a. paracentric inversion

b. deletion

c. pericentric inversion

d. duplication

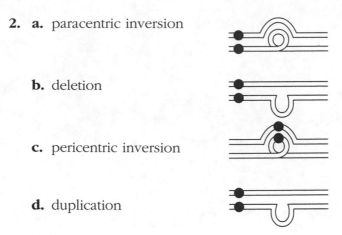

3. a. The products of crossing-over within the inversion will be inviable when the inversion is heterozygous. This paracentric inversion spans 25% of the region between the two loci and therefore will reduce the observed recombination between these genes by a similar percentage (i.e., 9%). The observed RF will be 27%.

b. When the inversion is homozygous, the products of crossing-over within the inversion will be viable, so the observed RF will be 36%.

4. The following represents the crosses that are described in this problem:

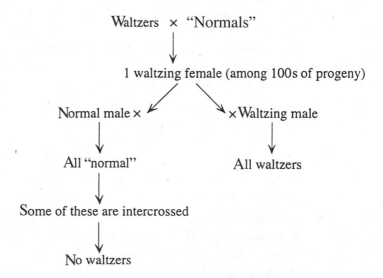

The single waltzing female that arose from a cross between waltzers and normals is expressing a recessive gene. It is possible that this represents a new "waltzer" mutation that was inherited from one of the "normal" mice, but given the cytological evidence (the presence of a shortened chromosome), it is more likely that this exceptional female inherited a deletion of the wild-type allele, which allowed expression of the mutant recessive phenotype.

When this exceptional female was mated to a waltzing male, all the progeny were waltzers; when mated to a normal male, all the progeny were normal. When some of these normal offspring were intercrossed, there were no progeny that were waltzers. If a "new" recessive waltzer allele had been inherited, all these "normal" progeny would have been w^+/w. Any intercross should have therefore produced 25% waltzers. On the other hand, if a deletion had occurred, half the progeny would be w^+/w and half would be $w^+/w^{deletion}$. If $w^+/w^{deletion}$ are intercrossed, 25% of the progeny would not develop (the homozygous deletion would likely be lethal), and no waltzers would be observed. This is consistent with the data.

5. This problem uses a known set of overlapping deletions to order a set of mutants. This is called **deletion mapping** and is based on the expression of the recessive mutant phenotype when heterozygous with a deletion of the corresponding allele on the other homolog. For example, mutants a, b, and c are all expressed when heterozygous with Del1. Thus it can be assumed that these genes are deleted in Del1. When these results are compared with the crosses with Del2 and it is discovered that these progeny are b^+, the location of gene b is mapped to the region deleted in Del1 that is not deleted in Del2. This logic can be applied in the following way:

Compare deletions 1 and 2: this places allele b more to the left than alleles a and c. The order is $b\,(a, c)$, where the parentheses indicate that the order is unknown.

Compare deletions 2 and 3: this places allele e more to the right than (a, c). The order is $b\,(a, c)\,e$.

Compare deletions 3 and 4: allele a is more to the left than c and e, and d is more to the right than e. The order is $b\,a\,c\,e\,d$.

Compare deletions 4 and 5: allele f is more to the right than d. The order is $b\,a\,c\,e\,d\,f$.

Allele	Band
b	1
a	2
c	3
e	4
d	5
f	6

6. The data suggest that one or both breakpoints of the inversion are located within an essential gene, causing a recessive lethal mutation.

7. **a.** The Sumatra chromosome contains a pericentric inversion when compared with the Borneo chromosome.

b.

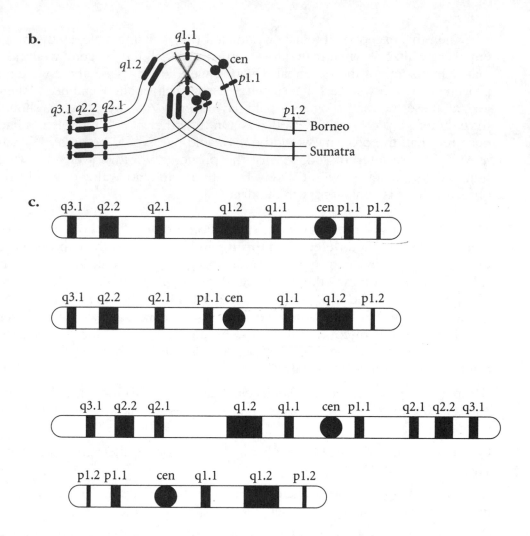

c.

d. Recall that all single crossovers within the inverted region will lead to four meiotic products: two that will be viable, nonrecombinant (parental) types and two that will be extremely unbalanced (most likely nonviable), recombinant types. In other words, if 30% of the meioses have a crossover in this region, 15% of the gametes will not lead to viable progeny. That means that 85% of the gametes should produce viable progeny.

8. *Unpacking the Problem*

   a. A "gene for tassel length" means that there is a gene with at least two alleles (T and t) that controls the length of the tassel. A "gene for rust resistance" means that there is a gene that determines whether the corn plant is resistant to a rust infection or not (R and r).

   b. The precise meaning of the allelic symbols for the two genes is irrelevant to solving the problem because what is being investigated is the distance between the two genes.

c. A **locus** is the specific position occupied by a gene on a chromosome. It is implied that gene loci are the same on both homologous chromosomes. The gene pair can consist of identical or different alleles.

d. Evidence that the two genes are normally on separate chromosomes would have come from previous experiments showing that the two genes independently assort during meiosis.

e. Routine crosses could consist of F_1 crosses, F_2 crosses, backcrosses, and testcrosses.

f. The genotype T/t ; R/r is a double heterozygote, or dihybrid, or F_1 genotype.

g. The pollen parent is the "male" parent that contributes to the pollen tube nucleus, the endosperm nucleus, and the progeny.

h. Testcrosses are crosses that involve a genotypically unknown and a homozygous recessive organism. They are used to reveal the complete genotype of the unknown organism and to study recombination during meiosis.

i. The breeder was expecting to observe 1 T/t ; R/r:1 T/t ; r/r:1 t/t ; R/r:1 t/t ; r/r.

j. Instead of a 1:1:1:1 ratio indicating independent assortment, the testcross indicated that the two genes were linked, with a genetic distance of $100\%(3 + 5)/210 = 3.8$ map units.

k. The equality and predominance of the first two classes indicate that the parentals were $T R/t r$.

l. The equality and lack of predominance of the second two classes indicate that they represent recombinants.

m. The gametes leading to this observation were:

46.7% $T R$	1.4% $T r$
49.5% $t r$	2.4% $t R$

n. 46.7% $T R$

   49.5% $t r$

o. 1.4% $T r$

   2.4% $t R$

p. $T r$ and $t R$

q. T and R are linked, as are t and r.

r. Two genes on separate chromosomes can become linked through a translocation.

s. One parent of the hybrid plant contained a translocation that linked the T and R alleles and the t and r alleles.

t. A corn cob is a structure that holds on its surface the progeny of the next generation.

u.

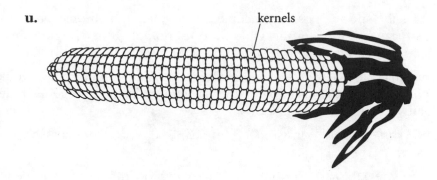

kernels

v.

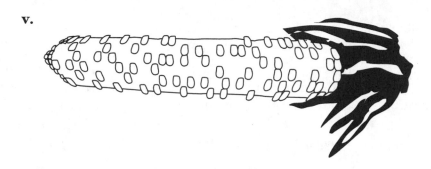

w. A kernel is one progeny on a corn cob.

x. Absence of half the kernels, or 50% aborted progeny (semisterility), could result from the random segregation of one normal with one translocated chromosome (T1 + N2 and T2 + N1) during meioses in a parent that is heterozygous for a reciprocal translocation.

y. Approximately 50% of the progeny died. It was the "female" that was heterozygous for the translocation.

Solution to the Problem

a. The progeny are not in the 1:1:1:1 ratio expected for independent assortment; instead, the data indicate close linkage. And, half the progeny did not develop, indicating semisterility.

b. These observations are best explained by a translocation that brought the two loci close together.

c. Parents: $T R/t\ r$ $\times$ t/t ; r/r

    Progeny: 98 $T R/t$; r

             104          $t\ r/t$; r

             3          $T\ r/t$; r

             5          $t R/t$; r

d. Assume a translocation heterozygote in coupling. If pairing is as diagrammed below, then you would observe the following:

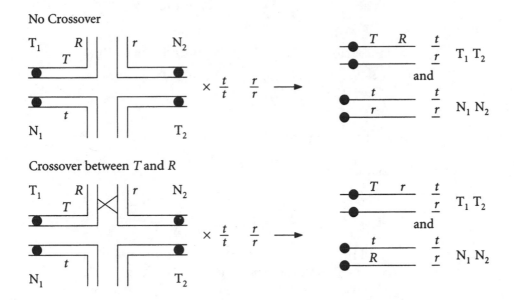

e. The two recombinant classes result from a recombination event followed by proper segregation of chromosomes, as diagrammed above.

9. The cross was

$$P \quad X^{e^+}/Y \text{ (irradiated)} \times X^e/X^e$$

$$F_1 \quad \text{Most } X^e/Y \text{ yellow males}$$

Two ? gray males

a. Gray male 1 was crossed with a yellow female, yielding yellow females and gray males, which is reversed sex linkage. If the e^+ allele was translocated to the Y chromosome, the gray male would be X^e/Y^{e^+} or gray. When crossed with yellow females, the results would be

$$X^e/Y^{e^+} \quad \text{Gray males}$$

$$X^e/X^e \quad \text{Yellow females}$$

b. Gray male 2 was crossed with a yellow female, yielding gray and yellow males and females in equal proportions. If the e^+ allele was translocated to an autosome, the progeny would be as below, where A indicates autosome:

$$P \quad A^{e^+}/A \ ; \ X^e/Y \times A/A \ ; \ X^e/X^e$$

$$F_1 \quad A^{e^+}/A \ ; \ X^e/X^e \quad \text{Gray female}$$

$$A^{e^+}/A \ ; \ X^e/Y \quad \text{Gray male}$$

$$A/A \ ; \ X^e/X^e \quad \text{Yellow female}$$

$$A/A \ ; \ X^e/Y \quad \text{Yellow male}$$

10. The break point can be treated as a gene with two "alleles," one for normal fertility and one for semisterility. The problem thus becomes a two-point cross.

Parentals	764	Semisterile *Pr*
	727	Normal *pr*
Recombinants	145	Semisterile *pr*
	186	Normal *Pr*
	1822	

$$100\%(145 + 186)/1822 = 18.17 \text{ m.u.}$$

11. Klinefelter syndrome   XXY male

Down syndrome   Trisomy 21

Turner syndrome   XO female

12. Create a hybrid by crossing the two plants and then double the chromosomes with a treatment that disrupts mitosis, such as colchicine treatment. Alternatively, diploid somatic cells from the two plants could be fused and then grown into plants through various culture techniques.

13. a.

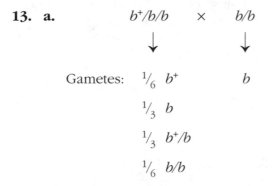

$$b^+/b/b \quad \times \quad b/b$$

Gametes: $\frac{1}{6}$ b^+ b

$\frac{1}{3}$ b

$\frac{1}{3}$ b^+/b

$\frac{1}{6}$ b/b

Among the progeny of this cross, the phenotypic ratio will be 1 wild type (b^+):1 b.

b. $b^+/b^+/b \quad \times \quad b/b$

Gametes: $\frac{1}{6}$ b b

$\frac{1}{3}$ b^+

$\frac{1}{3}$ b^+/b

$\frac{1}{6}$ b^+/b^+

Among the progeny of this cross, the phenotypic ratio will be 5 wild type (b^+):1 b.

c. $b^+/b^+/b$ × b^+/b

            ↓   ↓

Gametes: $\frac{1}{6}$ b $\frac{1}{2}$ b

            $\frac{1}{3}$ b^+ $\frac{1}{2}$ b^+

            $\frac{1}{3}$ b^+/b

            $\frac{1}{6}$ b^+/b^+

Among the progeny of this cross, the phenotypic ratio will be 11 wild type (b^+):1 b.

14. **a., b., c.** One of the parents of the woman with Turner syndrome (XO) must have been a carrier for color-blindness, an X-linked recessive disorder. Because her father has normal vision, she could not have obtained her only X from him. Therefore, nondisjunction occurred in her father. A sperm lacking a sex chromosome fertilized an egg with the X chromosome carrying the color-blindness allele. The nondisjunctive event could have occurred during either meiotic division.

    d. If the color-blind patient had Klinefelter syndrome (XXY), then both X's must carry the allele for color-blindness. Therefore, nondisjunction had to occur in the mother. Remember that during meiosis I, given no crossover between the gene and the centromere, allelic alternatives separate from each other. During meiosis II, identical alleles on sister chromatids separate. Therefore, the nondisjunctive event had to occur during meiosis II because both alleles are identical.

15. **a.** If a 6x were crossed with a 4x, the result would be 5x.

    b. Cross A/A with $a/a/a/a$ to obtain $A/a/a$.

    c. The easiest way is to expose the A/a^* plant cells to colchicine for one cell division. This will result in a doubling of chromosomes to yield $A/A/a^*/a^*$.

    d. Cross 6x ($a/a/a/a/a/a$) with 2x (A/A) to obtain $A/a/a/a$.

16. Type a: the extra chromosome must be from the mother. Because the chromosomes are identical, nondisjunction had to have occurred at M_{II}.

    Type b: The extra chromosome must be from the mother. Because the chromosomes are not identical, nondisjunction had to have occurred at M_I.

    Type c: The mother correctly contributed one chromosome, but the father did not contribute any chromosome 4. Therefore, nondisjunction occurred in the male during either meiotic division.

17. a. The cross is $P/P/p \times p/p$.

The gametes from the trisomic parent will occur in the following proportions:

$$\frac{1}{6} \quad p$$

$$\frac{2}{6} \quad P$$

$$\frac{1}{6} \quad P/P$$

$$\frac{2}{6} \quad P/p$$

Only gametes that are p can give rise to potato leaves, because potato is recessive. Therefore, the ratio of normal to potato will be 5:1.

b. If the gene is not on chromosome 6, there should be a 1:1 ratio of normal to potato.

18. The generalized cross is $A/A/A \times a/a$, from which $A/A/a$ progeny were selected. These progeny were crossed with a/a individuals, yielding the results given. Assume for a moment that each allele can be distinguished from the other, and let 1 = A, 2 = A, and 3 = a. The gametic combinations possible are

1–2 (A/A) and 3 (a)

1–3 (A/a) and 2 (A)

2–3 (A/a) and 1 (A)

Because only diploid progeny were examined in the cross with a/a, the progeny ratio should be 2 wild type:1 mutant if the gene is on the trisomic chromosome. With this in mind, the table indicates that y is on chromosome 1, cot is on chromosome 7, and h is on chromosome 10. Genes d and c do not map to any of these chromosomes.

19. a. Single crossovers between a gene and its centromere lead to a tetratype (second-division segregation). Thus, a total of 20% of the asci should show second-division segregation, and 80% will show first-division segregation. The following are representative asci:

$un3^+$ $ad3^+$	$un3^+$ $ad3^+$	$un3^+$ $ad3^+$	$un3^+$ $ad3$	$un3^+$ $ad3$
$un3^+$ $ad3^+$	$un3^+$ $ad3^+$	$un3^+$ $ad3^+$	$un3^+$ $ad3$	$un3^+$ $ad3$
$un3^+$ $ad3^+$	$un3^+$ $ad3$	$un3^+$ $ad3$	$un3^+$ $ad3^+$	$un3^+$ $ad3^+$
$un3^+$ $ad3^+$	$un3^+$ $ad3$	$un3^+$ $ad3$	$un3^+$ $ad3^+$	$un3^+$ $ad3^+$
$un3$ $ad3$	$un3$ $ad3^+$	$un3$ $ad3$	$un3$ $ad3^+$	$un3$ $ad3$
$un3$ $ad3$	$un3$ $ad3^+$	$un3$ $ad3$	$un3$ $ad3^+$	$un3$ $ad3$
$un3$ $ad3$	$un3$ $ad3$	$un3$ $ad3^+$	$un3$ $ad3$	$un3$ $ad3^+$
$un3$ $ad3$	$un3$ $ad3$	$un3$ $ad3^+$	$un3$ $ad3$	$un3$ $ad3^+$
80%	5%	5%	5%	5%

In all cases, the "upside-down" version would be equally likely.

b. The aborted spores could result from a crossing-over event within an inversion of the wild type compared with the standard strain. Crossing-over within heterozygous inversions leads to unbalanced chromosomes and nonviable spores. This could be tested by using the wild type from Hawaii in mapping experiments of other markers on chromosome 1 in crosses with the standard strain and looking for altered map distances.

20. Cross 1: Independent assortment of the two genes (expected for genes on separate chromosomes)

Cross 2: The two genes now appear to be linked (the observed RF is 1%); also, half of the progeny are inviable. These data suggest a reciprocal translocation occurred and both genes are very close to the breakpoints.

Cross 3: The viable spores are of two types: half contain the normal (nontranslocated chromosomes) and half contain the translocated chromosomes.

21. a. $a^+/a^+/a/a$ × $a/a/a/a$

↓   ↓

Gametes: $\frac{1}{6}$ a^+/a^+ a/a

$\frac{2}{3}$ a^+/a

$\frac{1}{6}$ a/a

Among the progeny of this cross, the phenotypic ratio will be 5 wild type (a^+):1 a.

b. $a^+/a/a/a$ × $a/a/a/a$

↓   ↓

Gametes: $\frac{1}{2}$ a^+/a a/a

$\frac{1}{2}$ a/a

Among the progeny of this cross, the phenotypic ratio will be 1 wild type (a^+):1 a.

c. $a^+/a/a/a$ × $a^+/a/a/a$

↓   ↓

Gametes: $\frac{1}{2}$ a^+/a $\frac{1}{2}$ a^+/a

$\frac{1}{2}$ a/a $\frac{1}{2}$ a/a

Among the progeny of this cross, the phenotypic ratio will be 3 wild type (a+):1 a.

d. $a^+/a^+/a/a$ × $a^+/a/a/a$

↓ ↓

Gametes: $\frac{1}{6}\ a^+/a^+$ $\frac{1}{2}\ a^+/a$

$\frac{2}{3}\ a^+/a$ $\frac{1}{2}\ a/a$

$\frac{1}{6}\ a/a$

Among the progeny of this cross, the phenotypic ratio will be 11 wild type ($a+$):1 a.

22. Consider the following table, in which L and S stand for 13 large and 13 small chromosomes, respectively:

Hybrid	Chromosomes
G. hirsutum × *G. thurberi* ·	S, S, L
G. hirsutum × *G. herbaceum*	S, L, L
G. thurberi × *G. herbaceum*	S, L

Each parent in the cross must contribute half its chromosomes to the hybrid offspring. It is known that *G. hirsutum* has twice as many chromosomes as the other two species. Furthermore, its chromosomes are composed of chromosomes donated by the other two species. Therefore, the genome of *G. hirsutum* must consist of one large and one small set of chromosomes. Once this is realized, the rest of the problem essentially solves itself. In the first hybrid, the genome of *G. thurberi* must consist of one set of small chromosomes. In the second hybrid, the genome of *G. herbaceum* must consist of one set of large chromosomes. The third hybrid confirms the conclusions reached from the first two hybrids.

The original parents must have had the following chromosome constitution:

G. hirsutum	26 large, 26 small
G. thurberi	26 small
G. herbaceum	26 large

G. hirsutum is a polyploid derivative of a cross between the two Old World species. This could easily be checked by looking at the chromosomes.

23. **a.** Loss of one X in the developing fetus after the two-cell stage.

   b. Nondisjunction leading to Klinefelter syndrome (XXY), followed by a nondisjunctive event in one cell for the Y chromosome after the two-cell stage, resulting in XX and XXYY.

   c. Nondisjunction of the X at the one-cell stage.

d. Fused XX and XY zygotes (from the separate fertilizations either of two eggs or of an egg and a polar body by one X-bearing and one Y-bearing sperm).

e. Nondisjunction of the X at the two-cell stage or later.

24. Cross 1: P $\quad b\,e^+/b\,e^+ \times b^+\,e/b^+\,e$

$\quad\quad\quad$ F$_1$ $\quad b^+\,e/b\,e^+$

Cross 2: P $\quad$ X/X ; $b^+\,e/b\,e^+ \times$ X/Y ; $b\,e/b\,e$

$\quad\quad\quad$ F$_1$ $\quad$ expect 1 $b\,e^+/b\,e$:1 $b^+\,e/b\,e$, X/X and X/Y

$\quad\quad\quad\quad$ one rare observed X/X ; $b^+\,e^+$

a. The common progeny are $b^+\,e/b\,e$ and $b\,e^+/b\,e$.

b. The rare female could have come from crossing-over, which would have resulted in a gamete that was $b^+\,e^+$. The rare female also could have come from nondisjunction that gave a gamete that was $b\,e^+/b^+\,e$. Such a gamete might give rise to viable progeny.

c. If the female had been wild type ($b^+\,e^+/b\,e$) as a result of crossing-over, her progeny would have been as follows:

Parental:	$b^+\,e^+/b\,e$	Wild type (common)
	$b\,e/b\,e$	Bent, eyeless (common)
Recombinant:	$b\,e^+/b\,e$	Bent (rare)
	$b^+\,e/b\,e$	Eyeless (rare)

These expected results are very far from what was observed, so the rare female was not the result of recombination.

$\quad$ If the female had been the product of nondisjunction ($b\,e^+/b^+\,e/b\,e$), her progeny when crossed to $b\,e/b\,e$ would be as follows:

$1/6$	$b^+\,e/b\,e$	Eyeless
$1/6$	$b\,e^+/b\,e/b\,e$	Bent
$1/6$	$b^+\,e/b\,e/b\,e$	Eyeless
$1/6$	$b\,e^+/b\,e$	Bent
$1/6$	$b\,e/b\,e$	Bent, eyeless
$1/6$	$b\,e^+/b^+\,e/b\,e$	Wild type

Overall, 2 bent:2 eyeless:1 bent eyeless:1 wild type

$\quad$ These results are in accord with the observed results, indicating that the female was a product of nondisjunction.

25. Recall that ascospores are haploid. The normal genotype associated with the phenotype of each spore is given below:

1	2	3
b^+f^+	bf^+	b^+f
b^+f^+	bf^+	b^+f
b^+f^+	abort	b^+f^+
b^+f^+	abort	b^+f^+
abort	b^+f	bf
abort	b^+f	bf
abort	b^+f	bf^+
abort	b^+f	bf^+

a. For the first ascus, the most reasonable explanation is that nondisjunction occurred at the first meiotic division. Second-division nondisjunction or chromosome loss are two explanations of the second ascus. Crossing-over best explains the third ascus.

b.

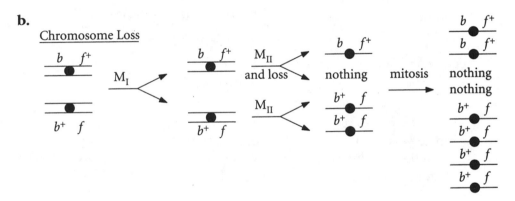

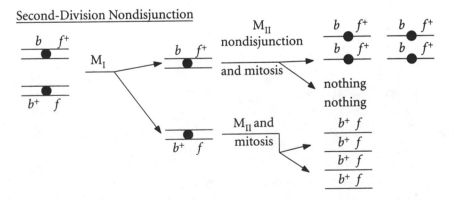

12 MUTATIONAL DISSECTION

1. All cells derived from the cell in which the reversion took place will now be w^+/w. Depending on when during development this took place, the petal will now be blue, either in part or in whole. Since the petal is part of the plant's soma, this reversion would not be inherited.

2. Starting with a yeast strain that is *pro-1*, plate the cells on medium lacking proline. Only those cells that are able to synthesize proline will form colonies. Most of these will be revertants; however, some will have second-site suppressors. (Treating the cells with a mutagen prior to plating them would significantly increase the yield.)

3. There are many ways to test a chemical for mutagenicity. For example, the text discusses a detection system for recessive somatic mutations in mice (see Figure 15-14). Mice bred to be heterozygous for seven genes involved in coat color are exposed to a potential mutagen by injecting it into the uterus of their pregnant mother. Any somatic mutation from wild type to mutant at one of the seven loci will result in a patch (or mutant sector) of differently colored fur. The number of mutant sectors later found on these chemically treated mice would be compared with the number found on genetically identical, but chemically untreated, mice (the control mice). (A proper control would expose the control mice to exactly the same experimental protocol except for the caffeine. This would include injection of whatever solvent was used into the uterus of their mother at the same developmental time as for the experimental mice.)

Much larger screens or selections could be done with fungi. For example, haploid *Neurospora* auxotrophic for the amino acid leucine could be exposed to caffeine and then plated onto minimal medium selecting for *leu*$^+$ colonies. Only those cells in which a reverse mutation (from *leu*$^-$ to *leu*$^+$) occurred would grow. Reversion rates of treated and untreated (control) cells would be compared to see if the caffeine was mutagenic. Alternatively, yeast cells could be exposed to caffeine and mutations in the gene *ade-3* could be scored and numbers compared between treated and untreated populations. (Mutations in this gene actually cause the yeast to be red instead of white, so large numbers of colonies can be screened rapidly.)

Although this question asks specifically for mutations in higher organisms, a rapid and widely used mutagen-detection system using bacteria was developed by Bruce Ames in the 1970s. Using a genetically modified bacterium (*Salmonella typhimurium*) that is auxotrophic for histidine and defective in DNA repair, the Ames test quickly ascertains the mutagenicity of various chemicals. Since the basic properties of DNA and mutation are the same in prokaryotes and eukaryotes, this test does have relevance. The Ames test has been further enhanced by using rat liver extracts to modify the tested chemicals to simulate human (and mammalian) metabolism. This is important because although the liver is responsible for most of the detoxification and metabolism of ingested chemicals (hence the connection between alcohol and liver disease!), some chemicals are modified in ways that actually make them toxic or mutagenic.

4. The commission was looking for induced recessive X-linked lethal mutations, which would show up as a shift in the sex ratio. A shift in the sex ratio is the first indication that a population has sustained lethal genetic damage. Other recessive mutations might have occurred, of course, but they would not be homozygous and therefore would go undetected. All dominant mutations would be immediately visible, unless they were lethal. If they were lethal, there would be lowered fertility, an increase in detected abortions, or both, but the sex ratio would not shift as dramatically.

5. The mutants can be categorized as follows:

   Mutant 1: an auxotrophic mutant

   Mutant 2: a non-nutritional, temperature-sensitive mutant

   Mutant 3: a leaky, auxotrophic mutant

   Mutant 4: a leaky, non-nutritional, temperature-sensitive mutant

   Mutant 5: a non-nutritional, temperature-sensitive, auxotrophic mutant

6. **a.** To select for a nerve mutation that blocks flying, place *Drosophila* at the bottom of a cage and place a poisoned food source at the top of the cage.

   b. Make antibodies against flagellar protein and expose mutagenized cultures to the antibodies.

   c. Do filtration through membranes with variously sized pores.

d. Screen visually.

e. Go to a large shopping mall and set up a rotating polarized disk. Ask the passersby to look through the disk for a free evaluation of their vision and their need for sunglasses. People with normal vision will see light with a constant intensity through the disk. Those with polarized vision will see alternating dark and light.

f. Set up a Y tube (a tube with a fork giving the choice of two pathways) and observe whether the flies or unicellular algae move to the light or the dark pathway.

g. Set up replica cultures and expose one of the two plates to low doses of UV.

7. The allele for NF must have arisen spontaneously in one of the parents' germ lines. Depending on when this mutation happened (the size of the mutant clone of germ-line tissue), their chance of having another affected child would range from 0 to 50% (the latter number if the entire germ line was mutant).

8. Phenocopying is the mimicking of a mutant phenotype by inactivating the gene product rather than the gene itself. It can be used regardless of how well-developed the genetic technology has been elaborated for a particular organism and can provide meaningful results when gene knockout or site-directed mutagenesis is not possible. Three methods of producing phenocopies are: the prevention of gene-specific translation by the introduction of anti-sense RNA (RNA complementary to the gene-specific mRNA); the introduction of double-stranded RNA with sequences homologous to part of a specific mRNA resulting in major reduction in the level of the mRNA (called double-stranded RNA interference); and inhibition of a specific protein through high-affinity binding of compound(s) identified through chemical genetics.

9. In cross-fertilizing species, F_3 individuals must be analyzed in order to recover homozygous autosomal recessive mutations after mutagenesis (see Figure 12-21b in the companion text). A balancer chromosome typically contains a dominant morphological marker and a series of inversions that prevent crossing over between it and its homolog. These features make mutant screens much more efficient, as any F_3 individual that lacks the balancer's dominant morphological marker will be homozygous for the homologous mutagenized chromosome (see Figure 12-22 in the companion text).

10. A forward mutation is any change away from the wild-type allele while any change back to the wild-type allele is called a reverse mutation.

11. Genetic selections permit highly efficient recovery of mutations, since individuals die if they lack a newly induced mutation affecting the phenotype in question. This allows for rapid identification of even very rare events; as only desired mutants grow, large numbers of organisms can be tested easily. Genetic screens are much less efficient (often similar to searching for a needle in haystack) but are much more flexible in allowing the investigator to recover mutations with essentially any kind of phenotype—even those that were unexpected!

12. *Unpacking the Problem*

a.

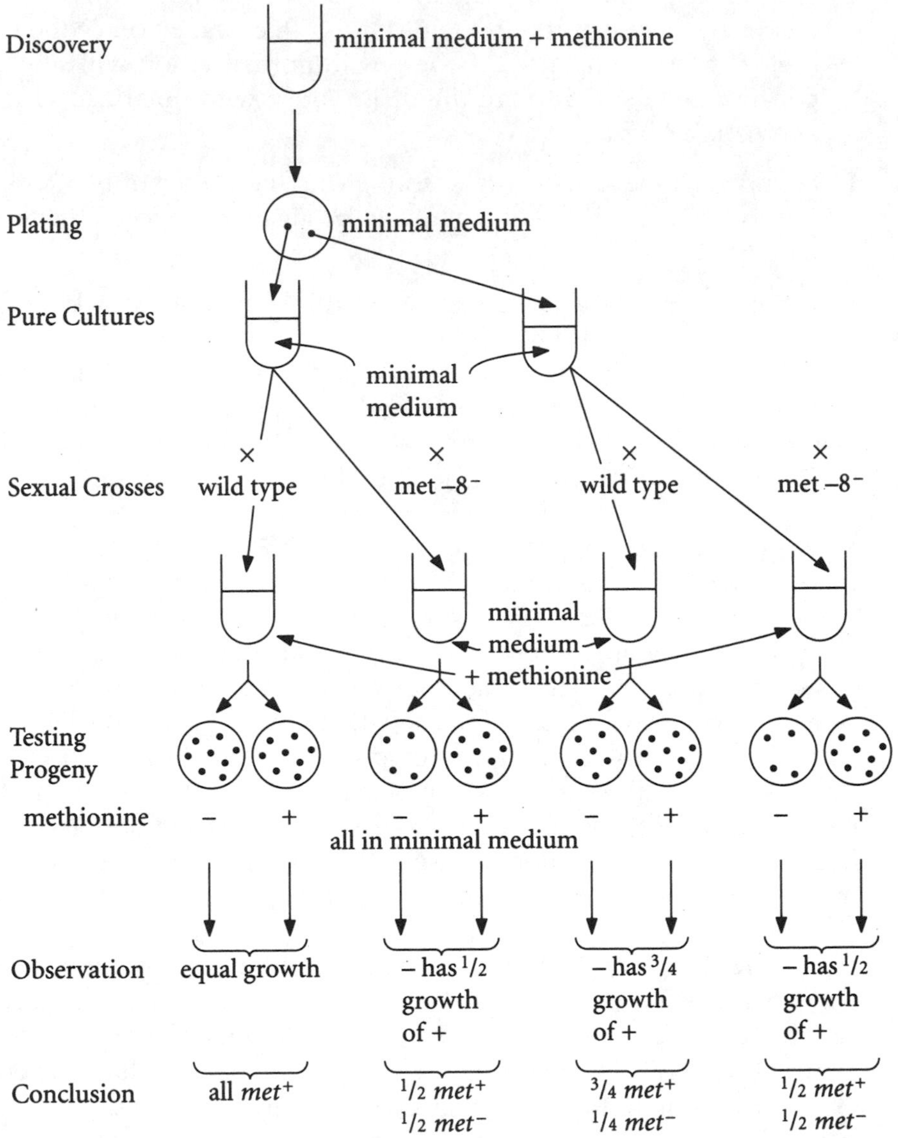
Discovery
minimal medium + methionine
Plating
minimal medium
Pure Cultures
minimal medium
Sexual Crosses
× wild type
× met –8⁻
× wild type
× met –8⁻
minimal medium + methionine
Testing Progeny
methionine
– +
– +
– +
– +
all in minimal medium
Observation
equal growth
– has 1/2 growth of +
– has 3/4 growth of +
– has 1/2 growth of +
Conclusion
all met⁺
1/2 met⁺ 1/2 met⁻
3/4 met⁺ 1/4 met⁻
1/2 met⁺ 1/2 met⁻

b. Haploid refers to possessing a single genome.

Auxotrophic means that an organism requires dietary provision of some substance that normally is not required by members of its species.

Methionine is an amino acid.

Asexual spores are a mode of propagation used by some species in which the spores are derived from an organism without a genetic contribution from another organism. Of necessity, the spores have the same number of genomes as the organism from which they are derived.

Prototrophic means that an organism does not have any special dietary requirements beyond those normal for the species.

A **colony** is a collection of cells or organisms all derived mitotically, from a single cell or organism and all possessing the same genotype.

A **mutation** is the process that generates alternative forms of genes, and it results in an inherited difference between parent and progeny.

c. The "8" in *met-8* refers to the eighth locus found that leads to a methionine requirement. It is unnecessary to know the specifics of the mutation in order to work the problem.

d. The following crosses were made in this problem:

> Cross 1: prototroph 1 × wild type
>
> Cross 2: prototroph 1 × backcross to *met-8*
>
> Cross 3: prototroph 2 × wild type
>
> Cross 4: prototroph 2 × backcross to *met-8*

e. Use *met-8** to indicate the prototroph derived from the *met-8* strain.

> Cross 1: *met-8** 1 × *met-8⁺*
>
> Cross 2: *met-8** I × *met-8*
>
> Cross 3: *met-8** 2 × *met-8⁺*
>
> Cross 4: *met-8** 2 × *met-8*

f. In this organism, asexual spores give rise to an organism that is capable of forming sexual spores following a mating. Therefore, the original mutation occurred in somatic tissue that subsequently gave rise to germinal tissue.

g. Because the trait being selected is the ability to grow in the absence of methionine, a reversion is being studied.

h. Only two revertants were observed because reversion occurs at a much lower frequency than forward mutation.

i. The millions of asexual spores did not grow because they required methionine and the medium used did not contain methionine.

j. A low percentage of the millions of spores that did not grow would be expected to have other mutations that rendered them incapable of growth. In addition, a low percentage would be expected to have chromosome abnormalities that would lead to death.

k. The wild type used in this experiment was prototrophic, by definition; that is, **wild type** refers to the norm for a species, which means "prototrophic."

l. It is highly unlikely that visual inspection could distinguish between wild type and prototrophic revertants.

m. One way to select for a *met-8* mutation is to grow a large number of spores on a medium that lacks methionine. Filtration will separate those spores capable of growth from those incapable of growth. Once spores have been isolated that are incapable of growth in a medium lacking methionine, they can be tested for a methionine mutation by plating them on medium containing methionine. If they are capable of growth on this second medium, they are methionine auxotrophs.

n. The starting auxotrophic spores were haploid. Both mitotic crossing-over and haploidization require diploids. Therefore, it is unlikely that either process is involved with producing the observed results.

o. Cross 1: *met-8* × wild type → 1 *met-8*:1 wild type

Cross 2: *met-8* × *met-8* → all *met-8*

Cross 3: wild type × wild type → all wild type

p. While the analysis could have been conducted using tetrad analysis, it is more likely that random selection of progeny was used.

13. **a., b.** Hypomorphic mutations result in less gene product activity than in the wild type. It would be expected that these would be more frequent among mutations caused by base substitution than frameshift mutagens, since the former will likely change just a single amino acid while the latter will alter the entire amino acid sequence coded downstream of the lesion. Amorphic mutations result in the complete loss of gene product activity. These would be expected to be more frequent among frameshift mutations.

14. Neomorphic mutations result in novel gene activity and are dominant. Reversion of the dominant phenotype will commonly be the result of introducing another mutation in the already mutant gene that now eliminates its function completely. Most gene knockout mutations are recessive so it is likely that most "revertants" will actually be recessive loss-of-function mutations.

15. **a.** No, because the mutational target size will vary considerably. Genes come in different physical sizes and that may play a role, but more difficult to calculate are issues such as what proportion of mutations make the gene product sufficiently abnormal or what proportion of non-protein-coding mutations alter the expression of the gene sufficiently to cause the desired mutant phenotype.

b. Not likely. The dorsal fin is a complex structure and it would be expected that more than five genes would contribute to its development and of course, although it represents a lot of work, 40 independently isolated mutations is not sufficient to "saturate" the genome for mutations.

Certainly, the original screen could be extended to search for more mutations and screening after mutagenesis with different mutagens would also be of use. Since some genes that effect fin development might also be required for other developmental programs or might otherwise be lethal to the organism when absent, screening for conditional alleles might also identify genes missed by nonconditional screens.

16. **a.** Foward genetics identifies heritable differences by their phenotypes and map locations and precedes the molecular analysis of the gene products. Reverse genetics starts with an identified protein or RNA and works towards mutating the gene that encodes it (and in the process, discovers the phenotype when the gene is mutated). Since you have known proteins and want to

determine the phenotypes associated with loss-of-function mutations in the genes that encoded them, a reverse genetic approach is the answer.

b. The two general approaches would be either directed mutations in the gene of interest or the generation of phenocopies of the mutant phenotype by inactivating the gene product rather than the gene itself.

17. Leaky mutants are mutants with an altered protein product that retains a low level of function. Enzyme activity may, for instance, be reduced rather than abolished by a mutation.

18. **a.** You are told that all mutants are simple Mendelian recessives, so you need not worry about mutations that map to the chloroplast's genome. To determine the number of genes involved, a complementation test (conducting pairwise crosses) is performed. All combinations complement (are mutant for different genes) except for 1×4. In this cross, the F_1 is still mutant, the mutations fail to complement and are in the same gene. Thus, 3 genes are represented.

b. As stated above, 1 and 4 map to the same gene. Of the other combinations, only 2 and 3 show linkage. In this case, the testcross of the 2×3 F_1 produces 10% wild-type progeny, not the expected 25% if the genes were unlinked. You can also use these data to determine the map distance between genes 2 and 3. The percentage of wild-type progeny from the testcross will be equal to half that of the recombinants (the other half will be mutant for both genes). Thus genes 2 and 3 are 20 map units apart.

c. 1×2: $\qquad m_1/m_1 ; +/+ \quad \times \quad +/+ ; m_2/m_2$

$\downarrow$

$m_1/+ ; m_2/+ \quad \times \quad m_1/m_1 ; m_2/m_2$ (testcross)

$\downarrow$

25% $m_1/m_1 ; m_2/+$

25% $m_1/+ ; m_2/m_2$

25% $m_1/m_1 ; m_2/m_2$

25% $m_1/+ ; m_2/+$ (wild type)

1×3: $\qquad m_1/m_1 ; +/+ \quad \times \quad +/+ ; m_3/m_3$

$\downarrow$

$m_1/+ ; m_3/+ \quad \times \quad m_1/m_1 ; m_3/m_3$ (testcross)

$\downarrow$

25% $m_1/m_1 ; m_3/+$

25% $m_1/+ ; m_3/m_3$

25% $m_1/m_1 ; m_3/m_3$

25% $m_1/+ ; m_3/+$ (wild type)

1×4: $\qquad m_1/m_1 \quad \times \quad m_4/m_4$

$\downarrow$

$m_1/m_4 \quad \times \quad m_1/m_1 \text{ or } m_4/m_4 \text{(testcross)}$

$\downarrow$

50% $\quad m_1/m_4$

50% $\quad m_1/m_1$

2×3: $\qquad m_2+/m_2+ \quad \times \quad +m_3/+m_3$

$\downarrow$

$m_2+/+m_3 \quad \times \quad m_2 m_3/m_2 m_3 \text{ (testcross)}$

$\downarrow$

40% $\quad m_2+/m_2 m_3$

40% $\quad +m_3/m_2 m_3$

10% $\quad m_2 m_3/m_2 m_3$

10% $\quad + +/m_2 m_3 \text{ (wild type)}$

2×4: as in 1×2

3×4: as in 1×2

19. 15% are essential gene functions (such as enzymes required for DNA replication or protein synthesis).

25% are auxotrophs (enzymes required for the synthesis of amino acids or the metabolism of sugars, etc.).

60% are redundant or pathways not tested (genes for histones, tubulin, ribosomal RNAs, etc., are present in multiple copies; the yeast may require many genes only under unique or special situations or in other ways that are not necessary for life in the "lab").

20. a., b. It is likely that the observed abnormalities are the result of mitotic recombination. A crossover between the genes of interest and the centromere in this case will lead to "twin spots" (the adjacent patches of stubby and ebony body observed). On the other hand, a crossover that occurs between the two genes will lead to "single spots" (the solitary patches of ebony). The position of the two genes with respect to the centromere determines the phenotype of the single spot. When recombination occurs in the region between the genes, the gene more distal to the centromere becomes homozygous while the more proximal gene remains heterozygous. In this problem, since the single patches are ebony, *e* is more distal than *s*. The other product of this event will appear normal and will not be detected among the other "normal" cells.

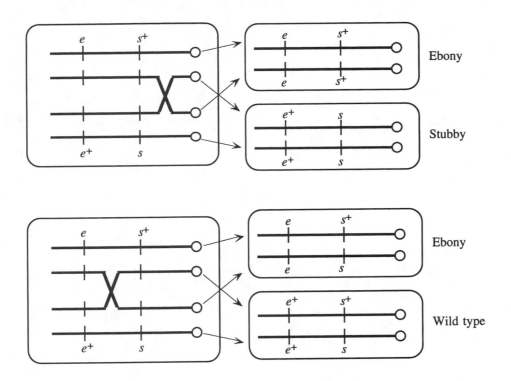

21. The cross is $trp^r \times trp^+$, where trp^r is the revertant. However, it might also be $trp^- ; su \times trp^+ ; su^+$, where su^+ has no effect on the trp gene and su is a suppressor of trp^-.

a. If the revertant is a precise reversal of the original change that produced the mutant allele, 100% of the progeny would be tryptophan independent.

b. If a suppressor mutation on a different chromosome is involved, then the cross is $trp^- ; su \times trp^+ ; su^+$. Independent assortment would lead to the following:

> 1 trp^- ; su tryptophan independent
>
> 1 trp^+ ; su^+ tryptophan independent
>
> 1 trp^- ; su^+ tryptophan dependent
>
> 1 trp^+ ; su tryptophan independent

c. If a suppressor mutation 24 map units from the trp locus occurred, then the cross is $trp^- \times su \times trp^+ \times su^+$ and the diploid intermediate is $trp^- \times su / trp^+ \times su^+$. The two parental types would occur 76% of the time, and the two recombinant types would occur 24% of the time. The progeny would be

> 38% $trp^- \times su$ tryptophan independent
>
> 38% $trp^+ \times su^+$ tryptophan independent
>
> 12% $trp^- \times su^+$ tryptophan dependent
>
> 12% $trp^+ \times su$ tryptophan independent

d. A common reason that a mutant is non-reverting is because it is the result of a deletion. However, since only "one" revertant was isolated from mutant B, it is possible that not enough cells were looked at to find an A revertant.

13

REGULATION OF GENE TRANSCRIPTION

1. The *I* gene determines the synthesis of a repressor molecule, which blocks expression of the *lac* operon and which is inactivated by the inducer. The presence of the repressor I^+ will be dominant to the absence of a repressor I^-. I^s mutants are unresponsive to an inducer. For this reason, the gene product cannot be stopped from interacting with the operator and blocking the *lac* operon. Therefore, I^s is dominant to I^+.

2. O^c mutants are changes in the DNA sequence of the operator that impair the binding of the *lac* repressor. Therefore, the *lac* operon associated with the O^c operator cannot be turned off. Because an operator controls only the genes on the same DNA strand, it is cis (on the same strand) and dominant (cannot be turned off).

3. **a.** You are told that *a*, *b*, and *c* represent *lacI*, *lacO*, and *lacZ*, but you do not know which is which. Both a^- and c^- have constitutive phenotypes (lines 1 and 2) and therefore must represent mutations in either the operator (*lacO*) or the repressor (*lacI*). b^- (line 3) shows no β-gal activity and by elimination must represent the *lacZ* gene.

   Mutations in the operator will be cis-dominant and will cause constitutive expression of the *lacZ* gene only if it's on the same chromosome. Line 6 has c^- on the same chromosome as b^+ but the phenotype is still inducible (owing to c^+ in trans). Line 7 has a^- on the same chromosome as b^+ and is constitutive even though the other chromosome is a^+. Therefore *a* is *lacO*, *c* is *lacI*, and *b* is *lacZ*.

b. Another way of labeling mutants of the operator is to denote that they lead to a constitutive phenotype; *lacO⁻* (or *a⁻*) can also be written as *lacOᶜ*. There are also mutations of the repressor that fail to bind inducer (allolactose) as opposed to fail to bind DNA. These two classes have quite different phenotypes and are distinguished by *lacIˢ* (fails to bind allolactose and leads to a dominant uninducible phenotype in the presence of a wild-type operator) and *lacI⁻* (fails to bind DNA and is recessive). It is possible that line 3, line 4, and line 7 have *lacIˢ* mutations (since dominance cannot be ascertained in a cell that is also *lacOᶜ*) but the other *c⁻* alleles must be *lacI⁻*.

4.

Part	β-Galactosidase		Permease	
	No lactose	Lactose	No lactose	Lactose
a	+	+	−	+
b	+	+	−	−
c	−	−	−	−
d	−	−	−	−
e	+	+	+	+
f	+	+	−	−
g	−	+	−	+

a. The *Oᶜ* mutation leads to the constitutive synthesis of ß-galactosidase because it is cis to a *lacZ⁺* gene, but the permease is inducible because the *lacY⁺* gene is cis to a wild-type operator.

b. The *lacP⁻* mutation prevents transcription so only the genes cis to *lacP⁺* will be transcribed. These genes are also cis to *Oᶜ* so the *lacZ⁺* gene is transcribed constitutively.

c. The *lacIˢ* is a trans-dominant mutation and prevents transcription from either operon.

d. Same as part c.

e. There is no functional repressor made (and one operator is mutant as well).

f. Same as part b.

g. Both operators are wild type and the one functional copy of *lacI* will direct the synthesis of enough repressor to control both operons.

5. A gene is turned off or inactivated by the "modulator" (usually called a *repressor*) in negative control, and the repressor must be removed for transcription to occur. A gene is turned on by the "modulator" (usually called an *activator*) in positive control, and the activator must be added or converted to an active form for transcription to occur.

6. The *lacY* gene produces a permease that transports lactose into the cell. A cell containing a *lacY*⁻ mutation cannot transport lactose into the cell, so β-galactosidase will not be induced.

7. Activation of gene expression by trans-acting factors occurs in both prokaryotes and eukaryotes. In both cases, the trans-acting factors interact with specific DNA sequences that control expression of cis genes.

    In prokaryotes, proteins bind to specific DNA sequences, which in turn regulate one or more downstream genes.

    In eukaryotes, highly conserved sequences such as CCAAT and various enhancers in conjunction with trans-acting binding proteins increase transcription controlled by the downstream TATA box promoter. Several proteins have been found that bind to the CCAAT sequence, upstream GC boxes, and the TATA sequence in *Drosophila*, yeast, and other organisms. Specifically, the Sp1 protein recognizes the upstream GC boxes of the SV40 promoter and many other genes; GCN4 and GAL4 proteins recognize upstream sequences in yeast; and many hormone receptors bind to specific sites on the DNA (e.g., estrogen complexed to its receptor binding to a sequence upstream of the ovalbumin gene in chicken oviduct cells). Additionally, the structure of some of these trans-acting DNA-binding proteins is quite similar to the structure of binding proteins seen in prokaryotes. Further, protein-protein interactions are important in both prokaryotes and eukaryotes. For the above reasons, eukaryotic regulation is now thought to be very close to the model for regulation of the bacterial *ara* operon.

8. Bacterial operons contain a promoter region that extends approximately 35 bases upstream of the site where transcription is initiated. Within this region is the promoter. Activators and repressors, both of which are trans-acting proteins that bind to the promoter region, regulate transcription of associated genes in cis only.

    The eukaryotic gene has the same basic organization. However, the promoter region is somewhat larger. Also, enhancers up to several thousand nucleotides upstream or downstream can influence the rate of transcription. A major difference is that eukaryotes have not been demonstrated to have polycistronic messages.

9. The term **epigenetic inheritance** is used to describe heritable alterations in which the DNA sequence itself is not changed. Paramutation and parental imprinting are two such examples.

10. The *araC* product has two conformations, which are determined by the presence and absence of arabinose. When it has bound arabinose, the *araC* product can bind to the initiator site (*araI*) and activate transcription. When it is not bound to arabinose, the *araC* product binds to both the initiator (*araI*) and the operator (*araO*) sites, forming a loop of the intermediary DNA. When both sites are bound to the *araC* product, transcription is inhibited. The *araC* product is trans-acting.

Many eukaryotic trans-acting protein factors also bind to promoters, enhancers, or both that are upstream from the protein-encoding gene. These factors are required for the initiation of transcription. Additionally, some bind to other proteins, such as RNA polymerase II, in order to initiate transcription. Like their counterparts in the *ara* operon, the eukaryotic trans-acting protein factors can bind DNA at two sites, with the intermediary DNA forming a loop between the binding sites.

11. Normally, the repressor searches for the operator by rapidly binding and dissociating from nonoperator sequences. Even for sequences that mimic the true operator, the dissociation time is only a few seconds or less. Therefore, it is easy for the repressor to find new operators as new strands of DNA are synthesized. However, when the affinity of the repressor for DNA and operator is increased, it takes too long for the repressor to dissociate from sequences on the chromosome that mimic the true operator, and as the cell divides and new operators are synthesized, the repressor never quite finds all of them in time, leading to a partial synthesis of β-galactosidase. This explains why in the absence of IPTG there is some elevated β-galactosidase synthesis. When IPTG binds to the repressors with increased affinity, it lowers the affinity back to that of the normal repressor (without IPTG bound). Then, the repressor can rapidly dissociate from sequences in the chromosome that mimic the operator and find the true operator. Thus, β-galactosidase is repressed in the presence of IPTG in strains with repressors that have greatly increased affinity for operator. In summary, because of a kinetic phenomenon, we see a reverse induction curve.

12. Construct a set of reporter genes with the promoter region, the introns, and the region 3′ to the transcription unit of the gene in question containing different alterations that do not disrupt transcription or processing. Use these reporter genes to make transgenic animals by germ-line transformation. Assay for expression of the reporter gene in various tissues and the kidney of both sexes.

13. If there is an operon governing both genes, then a frameshift mutation could cause the stop codon separating the two genes to be read as a sense codon. Therefore, the second gene product will be incorrect for almost all amino acids. However, there are no known polycistronic messages in eukaryotes. The alternative, and better, explanation is that both enzymatic functions are performed by the same gene product. Here, a frameshift mutation beyond the first function, carbamyl phosphate synthetase, will result in the second half of the protein molecule being nonfunctional.

14. Because very small amounts of the repressor are made, the system as a whole is quite responsive to changes in lactose concentration. In the heterodiploids, repressor tetramers may form by association of polypeptides encoded by I^- and I^+. The operator binding site binds two subunits at a time. Therefore, the repressors produced may reduce operator binding, which in turn would result in some expression of the *lac* genes in the absence of lactose.

15. The key to this question is to remember that *lacI* mutations will be trans-acting as they produce a protein product, and that *lacP* mutations (like *lacO* mutations) will be cis-acting as they are affecting a binding site for RNA polymerase. For the purposes of this problem, designate the uninducible *lacI* mutations as i^u mutations, and the uninducible *lacP* mutations as p^u mutations. There are quite a number of satisfactory genotypes which can serve to distinguish between the i^u and p^u mutations. Here are a few examples.

| | | **Enzyme activity** | |
Genotype		β-galactosidase	Permease
1	$i^u p^+ o^+ z^+ y^+$	absent	absent
2	$i^+ p^u o^+ z^+ y^+$	absent	absent
3	$i^u p^+ o^+ z^+ y^- / i^+ p^+ o^+ z^- y^+$	absent	absent
4	$i^+ p^u o^+ z^+ y^- / i^+ p^+ o^+ z^- y^+$	absent	inducible
5	$i^u p^+ o^c z^+ y^- / i^+ p^+ o^+ z^- y^+$	constitutive	absent
6	$i^+ p^u o^c z^+ y^- / i^+ p^+ o^+ z^- y^+$	absent	inducible

Genotypes 1 and 2 are simply symbolic restatements of the phenotypes of the uninducible mutations. Genotypes 3 and 4 are straight-forward tests to distinguish the cis-acting p^u mutations from the trans-acting i^u lesions. The results of genotype 3 reflect the expectation that i^u mutations would be trans-acting and dominant to i^+. This is expected because the i^u-encoded repressor protein molecules would be incapable of being inactivated by binding to inducer; the presence or absence of normal repressor protein is irrelevant. The results of genotype 4 on the other hand reflect the expectation that p^u mutations would only be cis-acting. Hence, any genes in cis to the p^u allele would be inactive, while any genes in cis to the normal p^+ allele would potentially be transcribed normally (if all other regulatory functions were normal). In a similar fashion, genotypes 5 and 6 distinguish the cis vs. trans action of i^u and p^u mutations. In genotype 5, i^u remains trans-dominant to i^+, but this dominance is overcome by the cis-acting o^c mutation (compare genotypes 3 and 5). In genotype 6, the presence of o^c is irrelevant, as it is in a *lac* operon which contains the p^u mutation preventing RNA polymerase binding (compare genotypes 4 and 6).

16. The *S* mutation is an alteration in *lacI* such that the repressor protein binds to the operator regardless of whether inducer is present or not. In other words, it is a mutation that inactivates the allosteric site that binds to inducer, while not affecting the ability of the repressor to bind to the operator site. The dominance of the *S* mutation is due to the binding of the mutant repressor even under circumstances when normal repressor does not bind to DNA (that is, in the presence of inducer). The constitutive reverse mutations that map to *lacI* are mutational events that inactivate the ability of this repressor to bind to the operator. The constitutive reverse mutations that map to the operator alter the operator DNA sequence such that it will not permit binding to any repressor molecules (wild type or mutant repressor).

17. **a.**

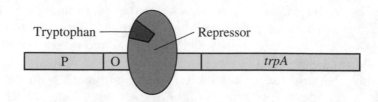

b. i. *trpA* is not synthesized in the presence of tryptophan; it is synthesized in the absence of tryptophan.

ii. *trpA* is not synthesized in the presence of tryptophan; it is synthesized in the absence of tryptophan. The second operon contains a *trpA⁻* mutation and is unable to make any active tryptophan synthetase enzyme. However, this operon contains a wild-type *trpR* gene, which encodes a functional repressor molecule. Since this gene product is a diffusible molecule (trans-acting), it will act on the other DNA molecule containing the mutant *trpR⁻* gene and bind to the operator on that molecule. Thus, *trpA* gene expression will be repressed when tryptophan binds the repressor, causing the repressor to bind the operator. In the absence of tryptophan, the repressor will not bind to either operator, and transcription will proceed.

iii. *trpA* is synthesized both in the presence and in the absence of tryptophan. Since the second operon contains a mutant *trpA⁻* gene, no functional tryptophan synthetase enzyme will be made from this DNA molecule. The first operon contains a wild-type *trpA* gene and will be responsible for the intracellular supply of tryptophan synthetase and ultimately, tryptophan. However, this operon contains a mutant (*trpO⁻*) operator region, which cannot be bound by the repressor molecules. Since the operator region is cis-acting and does not encode a diffusible gene product, the wild-type *trpO* gene on the other DNA molecule cannot substitute for the mutant operator. Therefore, the *trpA* gene will always be transcribed (constitutive) regardless of the levels of tryptophan in the cell.

18.

	Glucose	Lactose	Lactose + glucose
wild type	0	100	1
lacI⁻	1	100	1
lacIˢ	0	0	0
lacO⁻	1	100	1
crp⁻	0	1	1

lacI⁻—leads to absence of negative control by repressor binding but still under the positive control of CAP-cAMP binding

lacIˢ—since the repressor does not bind lactose but still binds DNA, transcription will be blocked under all conditions

lacO⁻—leads to absence of negative control by repressor binding but still under the positive control of CAP-cAMP binding *crp⁻*—leads to absence of positive control by CAP-cAMP binding but still under the negative control of repressor binding

19. a. D through J—the primary transcript will include all exons and introns

   b. E, G, I—all introns will be removed

   c. A, C, L—the promoter and enhancer regions will bind various transcription factors that may interact with RNA polymerase.

20. a. Clone A is activated by FGF, but B, C, and D are not. This indicates that the DNA binding site for this activation is located somewhere between the 3′ end of exon 1 [E1(f)] and the 5′ end of exon 3 [E3(f)]. This is the region of DNA found only in clone A.

   b. Cortisol represses transcription of clone A and B, but not C. (You do not expect any effect on D, the intact globin gene.) Comparing these clones indicates that the DNA site involved in this repression must be located in the 3′ region of E3(f) or the 3′ flanking sequences of this gene.

   c. Activation by EP is seen in clones C and D, but not B. (Again, you do not expect A, the intact *c-fos* gene to respond to EP.) This indicates that the DNA site involved in this activation must be localized to the 3′ side of E3(g) or the 3′ flanking regions of the globin gene.

14 From Gene to Phenotype

1. Complementation is the cooperation of the products of two or more genes to produce a nonmutant phenotype, while recombination is the exchange of DNA segments between chromosomes. Recombination can occur both within and between genes, while complementation occurs between gene products.

2. You are told that the cross of two erminette fowls results in 22 erminette, 14 black, and 12 pure white. Two facts are important: (1) the parents consist of only one phenotype, yet the offspring have three phenotypes, and (2) the progeny appear in an approximate ratio of 1:2:1. These facts should tell you immediately that you are dealing with a heterozygous × heterozygous cross involving one gene and that the erminette phenotype must be the heterozygous phenotype.

   When the heterozygote shows a different phenotype from either of the two homozygotes, the heterozygous phenotype results from incomplete dominance or codominance. Because two of the three phenotypes contain black, either fully or in an occasional feather, you might classify erminette as an instance of incomplete dominance because it is intermediate between fully black and fully white. Alternatively, because erminette has both black and white feathers, you might classify the phenotype as codominant. Your decision will rest on whether you look at the whole animal (incomplete dominance) or at individual feathers (codominance). This is yet another instance where what you conclude is determined by how you observe.

   To test the hypothesis that the erminette phenotype is a heterozygous phenotype, you could cross an erminette with either, or both, of the homozygotes. You should observe a 1:1 ratio in the progeny of both crosses.

3. From the cross $c^+/c^{cb} \times c^{cb}/c^b$ the progeny are

$1/4$	c^+/c^{cb}	Full color
$1/4$	c^+/c^b	Full color
$1/4$	c^{cb}/c^{cb}	Chinchilla
$1/4$	c^{cb}/c^b	Chinchilla

Thus, 50% of the progeny will be chinchilla.

4. a. The data indicate that there is a single gene with multiple alleles. The order of dominance is black > sepia > cream > albino.

Cross parents	Progeny	Conclusion
1 $b/a \times b/a$	3 $b/-$:1 a/a	Black is dominant to albino.
2 $b/s \times a/a$	1 b/a:1 s/a	Black is dominant to sepia; sepia is dominant to albino.
3 $c/a \times c/a$	3 $c/-$:1 a/a	Cream is dominant to albino.
4 $s/a \times c/a$	1 c/a:2 $s/-$:1 a/a	Sepia is dominant to cream.
5 $b/c \times a/a$	1 b/a:1 c/a	Black is dominant to cream.
6 $b/s \times c/-$	1 $b/-$:1 $s/-$	"$-$" can be c or a.
7 $b/s \times s/-$	1 b/s:1 $s/-$	"$-$" can be s, c, or a.
8 $b/c \times s/c$	1 s/c:2 $b/-$:1 c/c	"$-$" can be s or c.
9 $s/c \times s/c$	3 $s/-$:1 c/c	"$-$" can be s or c.
10 $c/a \times a/a$	1 c/a:1 a/a	Cream is dominant to albino.

b. The progeny of the cross $b/s \times b/c$ will be $3/4$ black ($1/4$ b/b, $1/4$ b/c, $1/4$ b/s):$1/4$ sepia (s/c).

5. Both codominance (=) and classical dominance (>) are present in the multiple allelic series for blood type: $A = B$, $A > O$, $B > O$.

Parents' phenotype	Parents' possible genotypes	Parents' possible children
a. AB × O	$A/B \times O/O$	A/O, B/O
b. A × O	A/A or $A/O \times O/O$	A/O, O/O
c. A × AB	A/A or $A/O \times A/B$	A/A, A/B, A/O, B/O
d. O × O	$O/O \times O/O$	O/O

The possible genotypes of the children are

Phenotype	Possible genotypes
O	*O/O*
A	*A/A, A/O*
B	*B/B, B/O*
AB	*A/B*

Using the assumption that each set of parents had one child, the following combinations are the only ones that will work as a solution:

Parents	Child
a. AB × O	B
b. A × O	A
c. A × AB	AB
d. O × O	O

6. **a.** The sex ratio is expected to be 1:1.

   b. The female parent was heterozygous for an X-linked recessive lethal allele. This would result in 50% fewer males than females.

   c. Half of the female progeny should be heterozygous for the lethal allele and half should be homozygous for the nonlethal allele. Individually mate the F_1 females and determine the sex ratio of their progeny.

7. Note that the F_2 are in a 9:6:1 ratio. This suggests a dihybrid cross in which *A/–* ; *b/b* has the same appearance as *a/a* ; *B/–*. Let the disc phenotype be the result of *A/–* ; *B/–* and the long phenotype be the result of *a/a* ; *b/b*. The crosses are

P	*A/A* ; *B/B* (disc) × *a/a* ; *b/b* (long)
F_1	*A/a* ; *B/b* (disc)
F_2	9 *A/–* ; *B/–* (disc)
	3 *a/a* ; *B/–* (sphere)
	3 *A/–* ; *b/b* (sphere)
	1 *a/a* ; *b/b* (long)

8. The suggestion from the data is that the two albino lines had mutations in two different genes. When the extracts from the two lines were placed in the same test tube, they were capable of producing color because the gene product of one line was capable of compensating for the absence of a gene product from the second line.

a. The most obvious control is to cross the two pure-breeding lines. The cross would be A/A ; $b/b \times a/a$; B/B. The progeny will be A/a ; B/b, and all should be reddish purple.

b. The most likely explanation is that the red pigment is produced by the action of at least two different gene products.

c. The genotypes of the two lines should be A/A ; b/b and a/a ; B/B.

d. The F_1 would all be pigmented, A/a ; B/b. This is an example of complementation. The mutants are defective for different genes. The F_2 would be

9	$A/–$; $B/–$	Pigmented
3	a/a ; $B/–$	White
3	$A/–$; b/b	White
1	a/a ; b/b	White

9. a. Intercrossing mutant strains that all share a common recessive phenotype is the basis of the complementation test. This test is designed to identify the number of different genes that can mutate to a particular phenotype. If the progeny of a given cross still express the mutant phenotype, the mutations fail to complement and are considered alleles of the same gene; if the progeny are wild type, the mutations complement and the two strains carry mutant alleles of separate genes.

b. There are three genes represented in these crosses. All crosses except 2×3 (or 3×2) complement and indicate that the strains are mutant for separate genes. Strains 2 and 3 fail to complement and are mutant for the same gene.

c. Let A and a represent alleles of gene 1; B and b represent alleles of gene 4; and c^2, c^3, and C represent alleles of gene 3.

> Line 1: a/a . B/B . C/C
>
> Line 2: A/A . B/B . c^2/c^2
>
> Line 3: A/A . B/B . c^3/c^3
>
> Line 4: A/A . b/b . C/C

	Cross	Genotype	Phenotype
F_1s	1×2	A/a . B/B . C/c^2	Wild type
	1×3	A/a . B/B . C/c^3	Wild type
	1×4	A/a . B/b . C/C	Wild type
	2×3	A/A . B/B . c^2/c^3	Mutant
	2×4	A/A . B/b . C/c^2	Wild type
	3×4	A/A . B/b . C/c^3	Wild type

d. With the exception that strains 2 and 3 fail to complement and therefore have mutations in the same gene, this test does not give evidence of linkage. To test linkage, the F₁s should be crossed to tester strains (homozygous recessive strains) and segregation of the mutant phenotype followed. If the genes are unlinked, for example, A/a ; B/b × a/a ; b/b, then 25% of the progeny will be wild type (A/a ; B/b) and 75% will be mutant (25% A/a ; b/b, 25% a/a ; B/b, and 25% a/a ; b/b). If the genes are linked ($a\,B/a\,B$ × $A\,b/A\,b$) then only one half of the recombinants (i.e., less than 25% of the total progeny) will be wild type ($A\,B/a\,b$).

e. No. All it tells you is that among these strains, there are three genes represented. If genetic dissection of leg coordination were desired, large screens for the mutant phenotype would be executed with the attempt to "saturate" the genome with mutations in all genes involved in the process.

10. a. Complementation refers to gene products within a cell, which is not what is happening here. Most likely, what is known as cross-feeding is occurring, whereby a product made by one strain diffuses to another strain and allows growth of the second strain. This is equivalent to supplementing the medium. Because cross-feeding seems to be taking place, the suggestion is that the strains are blocked at different points in the metabolic pathway.

b. For cross-feeding to occur, the growing strain must have a block that occurs earlier in the metabolic pathway than does the block in the strain from which it is obtaining the product for growth.

c. The *trpE* strain grows when cross-fed by either *trpD* or *trpB* but the converse is not true (placing *trpE* earlier in the pathway than either *trpD* or *trpB*), and *trpD* grows when cross-fed by *trpB* (placing *trpD* prior to *trpB*). This suggests that the metabolic pathway is

$$trpE \rightarrow trpD \rightarrow trpB$$

d. Without some tryptophan, no growth at all would occur, and the cells would not have lived long enough to produce a product that could diffuse.

11. a. This is an example where one phenotype in the parents gives rise to three phenotypes in the offspring. The "frizzle" is the heterozygous phenotype and shows incomplete dominance.

$$P \qquad A/a \text{ (frizzle)} \times A/a \text{ (frizzle)}$$

$$F_1 \qquad 1\ A/A \text{ (normal)}:2\ A/a \text{ (frizzle)}:1\ a/a \text{ (woolly)}$$

b. If A/A (normal) is crossed to a/a (woolly), all offspring will be A/a (frizzle).

12. a. The best explanation is that Marfan's syndrome is inherited as a dominant autosomal trait since roughly half of the children of all affected individuals also express the trait. If it were recessive, then all individuals marrying affected spouses would have to be heterozygous for an allele that when homozygous causes Marfan's.

b. The pedigree shows both pleiotropy (multiple affected traits) and variable expressivity (variable degree of expressed phenotype). Penetrance is the percentage of individuals with a specific genotype who express the associated phenotype. There is no evidence of decreased penetrance in this pedigree.

c. Pleiotropy indicates that the gene product is required in a number of different tissues, organs, or processes. When the gene is mutant, all tissues needing the gene product will be affected. Variable expressivity of a phenotype for a given genotype indicates modification by one or more other genes, random noise, or environmental effects.

13. It is possible to produce black offspring from two pure-breeding recessive albino parents if albinism results from mutations in either of two different genes. If the cross is designated

$$A/A \cdot b/b \times a/a \cdot B/B$$

then all the offspring would be

$$A/a \cdot B/b$$

and they would have a black phenotype because of complementation.

14. ***Unpacking the Problem***

a. The character being studied is petal color.

b. The wild-type phenotype is blue.

c. A variant is a phenotypic difference from wild type that is observed.

d. There are two variants: pink and white.

e. "In nature" means that the variants did not appear in laboratory stock and, instead, were found growing wild.

f. Possibly the variants appeared as a small patch or even a single plant within a larger patch of wild type.

g. Seeds would be grown to check the outcome from each cross.

h. Given that no sex linkage appears to exist (sex is not specified in parents or offspring), "blue × white" means the same as "white × blue." Similar results would be expected because the trait being studied appears to be autosomal.

i. The first two crosses show a 3:1 ratio in the F_2, suggesting the segregation of one gene. The third cross has a 9:4:3 ratio for the F_2, suggesting that two genes are segregating.

j. Blue is dominant to both white and pink.

k. *Complementation* refers to generation of wild-type progeny from the cross of two strains that are mutant in different genes.

l. The ability to make blue pigment requires two enzymes that are individually defective in the pink or white strains. The F_1 progeny of this cross is blue since each has inherited one nonmutant allele for both genes and can therefore produce both functional enzymes.

m. Blueness from a pink × white cross arises through complementation.

n. The following ratios are observed: 3:1, 9:4:3.

o. There are monohybrid ratios observed in the first two crosses.

p. There is a modified 9:3:3:1 ratio in the third cross.

q. A monohybrid ratio indicates that one gene is segregating, while a dihybrid ratio indicates that two genes are segregating.

r. 15:1, 12:3:1, 9:6:1, 9:4:3, 9:7

s. There is a modified dihybrid ratio in the third cross.

t. A modified dihybrid ratio most frequently indicates the interaction of two or more genes.

u. Recessive epistasis is indicated by the modified dihybrid ratio.

v.

w.

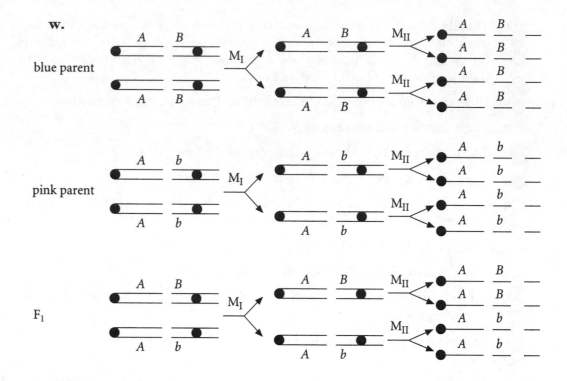

Solution to the Problem

a. Let A = wild type, a = white, B = wild type, and b = pink.

Cross 1:	P	Blue × white	A/A ; B/B × a/a ; B/B
	F_1	All blue	All A/a ; B/B
	F_2	3 blue:1 white	3 $A/-$; B/B:1 a/a ; B/B
Cross 2:	P	Blue × pink	A/A ; B/B × A/A ; b/b
	F_1	All blue	All A/A ; B/b
	F_2	3 blue:1 pink	3 A/A ; $B/-$:1 A/A ; b/b
Cross 3:	P	Pink × white	A/A ; b/b × a/a ; B/B
	F_1	All blue	All A/a ; B/b
	F_2	9 blue	9 $A/-$; $B/-$
		4 white	3 a/a ; $B/-$:1 a/a ; b/b
		3 pink	3 $A/-$; b/b

When the allele a is homozygous, the expression of allele B or b is blocked or masked. The white phenotype is epistatic to the pigmented phenotypes. It is likely that the product of the A gene produces an intermediate that is then modified by the product of the B gene. If the plant is a/a, this intermediate is not made and the phenotype of the plant is the same regardless of the ability to produce functional B product.

b. The cross is

$$F_2 \quad \text{Blue} \times \text{white}$$

$$F_3 \quad {}^3/_8 \text{ blue}$$

$$\quad\quad {}^1/_8 \text{ pink}$$

$$\quad\quad {}^4/_8 \text{ white}$$

Begin by writing as much of each genotype as can be assumed:

$$F_2 \quad A/- \; ; B/- \; \times \; a/a \; ; -/-$$

$$F_3 \quad {}^3/_8 \; A/- \; ; B/-$$

$$\quad\quad {}^1/_8 \; A/- \; ; b/b$$

$$\quad\quad {}^4/_8 \; a/a \; ; -/-$$

Notice that both a/a and b/b appear in the F_3 progeny. In order for these homozygous recessives to occur, each parent must have at least one a and one b. Using this information, the cross becomes

$$F_2 \quad A/a \; ; B/b \; \times \; a/a \; ; b/-$$

$$F_3 \quad {}^3/_8 \; \text{A}/a \; ; B/b$$

$$\quad\quad {}^1/_8 \; A/a \; ; b/b$$

$$\quad\quad {}^4/_8 \; a/a \; ; b/-$$

The only remaining question is whether the white parent was homozygous recessive, b/b, or heterozygous, B/b. If the white parent had been homozygous recessive, then the cross would have been a testcross of the blue parent, and the progeny ratio would have been 1 blue:1 pink:2 white, or 1 $A/a \; ; B/b$:1 $A/a \; ; b/b$:1 $a/a \; ; B/b$:1 $a/a \; ; b/b$. This was not observed. Therefore, the white parent had to have been heterozygous, and the F_2 cross was $A/a \; ; B/b \times a/a \; ; B/b$.

15. a., b. Crosses 1–3 show a 3:1 ratio, indicating that brown, black, and yellow are all alleles of one gene. Crosses 4–6 show a modified 9:3:3:1 ratio, indicating that at least two genes are involved. Those crosses also indicate that the presence of color is dominant to its absence. Furthermore, epistasis must be involved for there to be a modified 9:3:3:1 ratio.

By looking at the F_1 of crosses 1–3, the following allelic dominance relationships can be seen easily: black > brown > yellow. Arbitrarily assign the following genotypes for homozygotes: B^l/B^l = black, B^r/B^r = brown, B^y/B^y = yellow.

By looking at the F_2 of crosses 4–6, a white phenotype is composed of two categories: the double homozygote and one class of the mixed homozygote/heterozygote. Let lack of color be caused by c/c. Color will therefore be $C/-$.

Parents	F_1	F_2
1 B^r/B^r ; C/C × B^y/B^y ; C/C	B^r/B^y ; C/C	3 $B^r/-$; C/C:1 B^y/B^y ; C/C
2 B^l/B^l ; C/C × B^r/B^r ; C/C	B^l/B^r ; C/C	3 $B^l/-$; C/C:1 B^r/B^r ; C/C
3 B^l/B^l ; C/C × B^y/B^y ; C/C	B^l/B^y ; C/C	3 $B^l/-$; C/C:1 B^y/B^y ; C/C
4 B^l/B^l ; c/c × B^y/B^y ; C/C	B^l/B^y ; C/c	9 $B^l/-$; $C/-$:3 B^y/B^y ; $C/-$:3 $B^l/-$; c/c:1 B^y/B^y ; c/c
5 B^l/B^l ; c/c × B^r/B^r ; C/C	B^l/B^r ; C/c	9 $B^l/-$; $C/-$:3 B^r/B^r ; $C/-$:3 $B^l/-$; c/c:1 B^r/B^r ; c/c
6 B^l/B^l ; C/C × B^y/B^y ; c/c	B^l/B^y ; C/c	9 $B^l/-$; $C/-$:3 B^y/B^y ; $C/-$:3 $B^l/-$; c/c:1 B^y/B^y ; c/c

c. The following biochemical pathway is suggested by the data:

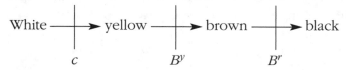

$$\text{White} \xrightarrow{\quad c \quad} \text{yellow} \xrightarrow{\quad B^y \quad} \text{brown} \xrightarrow{\quad B^r \quad} \text{black}$$

16. To solve this problem, first restate the information.

$A/-$	Yellow	$A/-$; $R/-$	Gray
$R/-$	Black	a/a ; r/r	White

The cross is gray × yellow, or $A/-$; $R/-$ × $A/-$; r/r. The F_1 progeny are

$3/_8$ yellow $\qquad$ $1/_8$ black

$3/_8$ gray $\qquad$ $1/_8$ white

For white progeny, both parents must carry an r and an a allele. Now the cross can be rewritten as:

$$A/a \ ; R/r \ \times \ A/a \ ; r/r$$

17. a. The stated cross is

P $\qquad$ Single-combed (r/r ; p/p) × walnut-combed (R/R ; P/P)

$\qquad\qquad$ F_1 $\qquad$ R/r ; P/p $\qquad$ Walnut

$\qquad\qquad$ F_2 $\qquad$ 9 $R/-$; $P/-$ $\qquad$ Walnut

$\qquad\qquad\qquad\qquad$ 3 r/r ; $P/-$ $\qquad$ Pea

$\qquad\qquad\qquad\qquad$ 3 $R/-$; p/p $\qquad$ Rose

$\qquad\qquad\qquad\qquad$ 1 r/r ; p/p $\qquad$ Single

b. The stated cross is

P   Walnut-combed × rose-combed

and the F$_1$ progeny are

Phenotypes		Possible genotypes
$^3/_8$	rose	$R/-$; p/p
$^3/_8$	walnut	$R/-$; $P/-$
$^1/_8$	pea	r/r ; $P/-$
$^1/_8$	single	r/r ; p/p

The 3 $R/-$:1 r/r ratio indicates that the parents were heterozygous for the R gene. The 1 $P/-$:1 p/p ratio indicates a testcross for this gene. Therefore, the parents were R/r ; P/p and R/r ; p/p.

c. The stated cross is

P   Walnut-combed × rose-combed

F$_1$ Walnut ($R/-$; P/p)

To get this result, one of the parents must be homozygous R, but both need not be, and the walnut parent must be homozygous P/P.

d. The following genotypes produce the walnut phenotype

R/R ; P/P, R/r ; P/P, R/R ; P/p, R/r ; P/p

18. a. This type of gene interaction is called *epistasis*. The phenotype of e/e is epistatic to the phenotypes of $B/-$ or b/b.

b. The progeny of generation I have all possible phenotypes. Progeny II-3 is beige (e/e), so both parents must be heterozygous E/e. Progeny II-4 is brown (b/b), so both parents must also be heterozygous B/b. Progeny III-3 and III-5 are brown, so II-2 and II-5 must be B/b. Progeny III-2 and III-7 are beige (e/e), so all their parents must be E/e.

The following are the inferred genotypes:

I   1 ($B/b\ E/e$)   2 ($B/b\ E/e$)

II   1 ($b/b\ E/e$)   2 ($B/b\ E/e$)   3 ($-/-\ e/e$)   4 ($b/b\ E/-$)   5 ($B/b\ E/e$)
         6 ($b/b\ E/e$)

III   1 ($B/b\ E/-$)   2 ($-/b\ e/e$)   3 ($b/b\ E/-$)   4 ($B/b\ E/-$)   5 ($b/b\ E/-$)
         6 ($B/b\ E/-$)   7 ($-/b\ e/e$)

19. a. Note that blue is always present, indicating E/E (blue) in both parents. Because of the ratios that are observed, neither C nor D is varying. In this case, the gene pairs that are involved are A/a and B/b. The parents are A/A ; b/b × a/a ; B/B or A/A ; B/B × a/a ; b/b.

The F_1 are A/a ; B/b and the F_2 are

9	$A/-$; $B/-$	Blue + red, or purple
3	$A/-$; b/b	Blue + yellow, or green
3	a/a ; $B/-$	Blue + white$_2$, or blue
1	a/a ; b/b	Blue + white$_2$, or blue

b. Blue is not always present, indicating E/e in the F_1. Because green never appears, the F_1 must be B/B . C/C . D/D. The parents are A/A ; e/e $\times$ a/a ; E/E or A/A ; E/E $\times$ a/a ; e/e.

The F_1 are A/a ; E/e and the F_2 are

9	$A/-$; $E/-$	Red + blue, or purple
3	$A/-$; e/e	Red + white$_1$, or red
3	a/a ; $E/-$	White$_2$ + blue, or blue
1	a/a ; e/e	White$_2$ + white$_1$, or white

c. Blue is always present, indicating that the F_1 is E/E. No green appears, indicating that the F_1 is also B/B. The two genes involved are A and D. The parents are A/A ; d/d $\times$ a/a ; D/D or A/A ; D/D $\times$ a/a ; d/d.

The F_1 are A/a ; D/d and the F_2 are

9	$A/-$; $D/-$	Blue + red + white$_4$, or purple
3	$A/-$; d/d	Blue + red, or purple
3	a/a ; $D/-$	Blue + white$_2$ + white$_4$, or blue
1	a/a ; d/d	White$_2$ + blue + red, or purple

d. The presence of yellow indicates b/b ; e/e in the F_2. Therefore, the parents are B/B ; e/e $\times$ b/b ; E/E or B/B ; E/E $\times$ b/b ; e/e.

The F_1 are B/b ; E/e and the F_2 are

9	$B/-$; $E/-$	Red + blue, or purple
3	$B/-$; e/e	Red + white$_1$, or red
3	b/b ; $E/-$	Yellow + blue, or green
1	b/b ; e/e	Yellow + white$_1$, or yellow

e. Mutations in D suppress mutations in A.

f. Recessive alleles of A will be epistatic to mutations in B.

20. a. The trait is recessive (parents without the trait have children with the trait) and autosomal (daughters can inherit the trait from unaffected fathers). Looking to generation III, there is also evidence that there are two different genes that when defective result in deaf-mutism.

Assuming that one gene has alleles *A* and *a* and the other has *B* and *b*, the following genotypes can be inferred:

I-1 and I-2	*A/a* ; *B/B*	I-3 and I-4	*A/A* ; *B/b*
II-(1, 3, 4, 5, 6)	*A/–* ; *B/B*	II- (9, 10, 12, 13, 14, 15)	*A/A* ; *B/–*
II-2 and II-7	*a/a* ; *B/B*	II-8 and II-11	*A/A* ; *b/b*

b. Generation III shows complementation. All are *A/a* ; *B/b*.

21. a. The first impression from the pedigree is that the gene causing blue sclera and brittle bones is pleiotropic with variable expressivity. If two genes were involved, it would be highly unlikely that all people with brittle bones also had blue sclera.

b. Sons and daughters inherit from affected fathers so the allele appears to be autosomal.

c. The trait appears to be inherited as a dominant but with incomplete penetrance. For the trait to be recessive, many of the nonrelated individuals marrying into the pedigree would have to be heterozygous (e.g., I-1, I-3, II-8, II-11). Individuals II-4, II-14, III-2, and III-14 have descendants with the disorder although they do not themselves express the disorder. Therefore, $^1/_5$ people that can be inferred to carry the gene, do not express the trait. That is 80% penetrance. (Penetrance could be significantly less than that since many possible carriers have no shown progeny.) The pedigree also exhibits variable expressivity. Of the 16 individuals who have blue sclera, 10 do not have brittle bones. Usually, expressivity is put in terms of none, variable, and highly variable, rather than expressed as percentages.

22. a. The first two crosses indicate that wild type is dominant to both platinum and aleutian. The third cross indicates that two genes are involved rather than one gene with multiple alleles because a 9:3:3:1 ratio is observed.

Let platinum be *a*, aleutian be *b*, and wild type be *A/–* ; *B/–*.

Cross 1:	P	*A/A* ; *B/B* × *a/a* ; *B/B*	Wild type × platinum
	F$_1$	*A/a* ; *B/B*	All wild type
	F$_2$	3 *A/–* ; *B/B*:1 *a/a* ; *B/B*	3 wild type:1 platinum
Cross 2:	P	*A/A* ; *B/B* × *A/A* ; *b/b*	Wild type × aleutian
	F$_1$	*A/A* ; *B/b*	All wild type
	F$_2$	3 *A/A* ; *B/–*:1 *A/A* ; *b/b*	3 wild type:1 aleutian
Cross 3:	P	*a/a* ; *B/B* × *A/A* ; *b/b*	Platinum × aleutian
	F$_1$	*A/a* ; *B/b*	All wild type
	F$_2$	9 *A/–* ; *B/–*	Wild type
		3 *A/–* ; *b/b*	Aleutian
		3 *a/a* ; *B/–*	Platinum
		1 *a/a* ; *b/b*	Sapphire

b. Sapphire × platinum

P	a/a ; b/b × a/a ; B/B	
F_1	a/a ; B/b	Platinum
F_2	3 a/a ; $B/-$	Platinum
	1 a/a ; b/b	Sapphire

Sapphire × aleutian

	a/a ; b/b × A/A ; b/b	
	A/a ; b/b	Aleutian
	3 $A/-$; b/b	Aleutian
	1 a/a ; b/b	Sapphire

23. a. The first experiment is a complementation test. This test is designed to identify the number of different genes involved in the analysis. In this problem,
if the heterokaryon still cannot grow in the absence of leucine, the mutations fail to complement and are considered alleles of the same gene; if the
heterokaryons grow, the mutations complement and the two strains carry
mutant alleles of separate genes.

The second experiment is a recombination test to indicate whether the
genes are linked. If the genes are unlinked, for example, A/a ; B/b × a/a ;
b/b, then 25% of the progeny will be wild type (A/a ; B/b) and 75% will be
mutant (25% A/a ; b/b, 25% a/a ; B/b, and 25% a/a ; b/b). If the genes are
linked ($a\ B/a\ B$ × $A\ b/A\ b$), then only one half of the recombinants (i.e., less
than 25% of the total progeny) will be wild type ($A\ B/a\ b$).

b. For all heterokaryons except those from a × e, growth in the absence of
leucine occurs. This indicates that a and e are mutations in the same gene
and that b, c, and d are mutations in separate genes. In other words, four
genes are being analyzed.

c. In the second experiment, if the genes are unlinked, 25% of the progeny
should be leucine-independent. The frequency of prototrophs is approximately 25% in all pairwise crosses except for a × e (0 prototrophic progeny)
and b × d (2 prototrophic progeny). From the complementation test, it is
already known that a and e are in the same gene. This test now indicates
that b and d are linked and that the RF = 100% × 4/500 = 0.8 map units.
(Remember, the prototrophs represent only one half of the recombinants.
The other recombinants are double mutants that will not be detected.)

d.

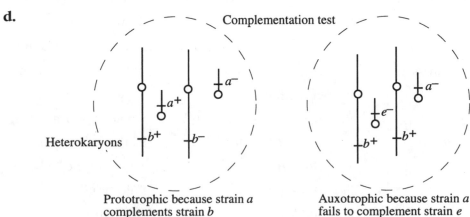

Complementation test

Heterokaryons

Prototrophic because strain a
complements strain b

Auxotrophic because strain a
fails to complement strain e

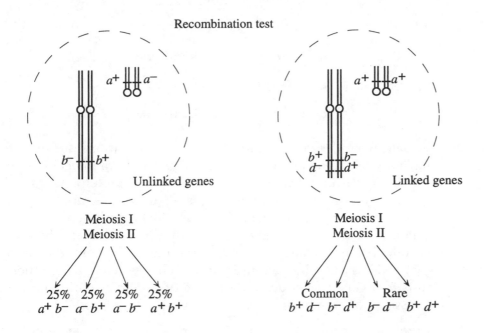

24. **a.** There are 10 mutants that fall into three complementation groups. Mutants 1, 2, 7 fail to complement; 3, 5, 8 fail to complement; and 4, 6, 9, 10 fail to complement. There are three genes indicated by this complementation test.

    b. There is evidence of a biochemical pathway. If a mutant grows on a supplement, the defect is earlier in the pathway; if a mutant fails to grow when supplemented, the defect is later in the pathway. Mutant 1 grows on SAICAR but not CAIR or AIR. The step blocked in mutant 1 is before SAICAR but after CAIR and AIR. Similarly, the step blocked in mutant 3 is after SAICAR, CAIR, and AIR, and the step blocked in mutant 4 is before CAIR and SAICAR but after AIR. The indicated pathway is

    Where a mutant is blocked is indicated by the vertical line through the arrow.

    c. Ten mutations that by complementation represent three genes involved in the metabolism of adenine are being studied. The recombination test indicates that the genes represented by mutants 1 and 3 are linked. The gene represented by mutant 4 is not linked to the other two. For unlinked genes, approximately 25% of the progeny should be prototrophic, and this is true for crosses 1×4 and 3×4. Cross 1×3 (and 3×1) showed 11 prototrophs out of a possible 2000 progeny. This represents half of the possible recombinants (the other half being doubly mutant) and indicates an RF = 100% $\times$ 22/2000 = 1.1%.

25. The cross is $e^+/e \; ; \; r^+/r \times e^+/e \; ; \; r^+/r$. For wild-type function, a functional regulatory protein and a functional gene encoding the enzyme are required. The progeny are

$9/16$	$e^+/-$; $r^+/-$	Have enzyme
$3/16$	$e^+/-$; r/r	Lack enzyme
$3/16$	e/e ; $r^+/-$	Lack enzyme
$1/16$	e/e ; r/r	Lack enzyme

26. **a.** The data from the crosses indicate that the mutations are in different, unlinked genes and both are recessive. The data from the gel indicate that one mutation is in the gene that codes for the P protein. As a result, the mutant protein is truncated (smaller) and runs more rapidly in the gel. Possible mutations giving this result might include nonsense, frameshifts, deletions, or mutations leading to altered splicing. (Alternatively, you could conclude that the mutation causes a larger, but still nonfunctional P protein to be made as a result of an insertion or splicing defect.) The other mutation is in a gene that codes for a regulatory protein required for P gene expression. When mutant, P expression is greatly reduced.

    b. Assume p^+ = normal P protein; r^+ = normal regulatory protein; both are required for normal function.

Lane 1:	p^+/p^+ ; $r^+/-$
Lane 2:	p^+/p^+ ; r/r
Lane 3:	p^+/p ; $r^+/-$
Lane 4:	p^+/p ; r/r
Lane 5:	p/p ; $r^+/-$
Lane 6:	p/p ; r/r

    c. Type 4 is p^+/p ; r/r, $1/2$ of the progeny will be p^+/p, and $1/4$ of the progeny will be independently r/r, so $1/2 \times 1/4 = 1/8$.

    d. Lane 1 (p^+/p^+ ; $r^+/-$) and lane 3 (p^+/p ; $r^+/-$) will be phenotypically wild type.

    e. Parents, p/p ; r^+/r^+ $\times$ p^+/p^+ ; r/r will look like lane 5 and lane 2, respectively.

    F_1, p^+/p ; r^+/r will look like lane 3.

27. The three mutations are recessive. Mutant line 1 and line 2 are mutations in the same gene (they do not complement) and mutant line 1 and line 3 are mutations in different genes (they do complement). Assume line 1 is m^1/m^1; line 2 is m^2/m^2; and line 3 is m^3/m^3.

    Line 1: $m^1/m^1 \times +/+ \rightarrow$ all $m^1/+ \rightarrow$ self $\rightarrow 3/4 \ +/-$:$1/4 \ m^1/m^1$

    Line 2: $m^2/m^2 \times +/+ \rightarrow$ all $m^2/+ \rightarrow$ self $\rightarrow 3/4 \ +/-$:$1/4 \ m^2/m^2$

    Line 3: $m^3/m^3 \times +/+ \rightarrow$ all $m^3/+ \rightarrow$ self $\rightarrow 3/4 \ +/-$:$1/4 \ m^3/m^3$

    Line 1 $\times$ line 2   Line 1 $\times$ line 3

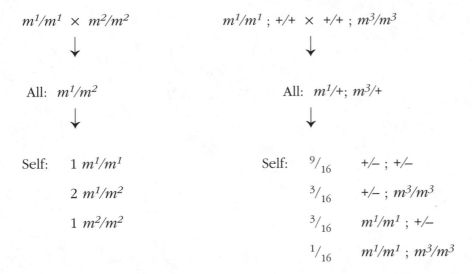

$m^1/m^1 \times m^2/m^2$

$\downarrow$

All: m^1/m^2

$\downarrow$

Self: 1 m^1/m^1

2 m^1/m^2

1 m^2/m^2

$m^1/m^1 ; +/+ \times +/+ ; m^3/m^3$

$\downarrow$

All: $m^1/+; m^3/+$

$\downarrow$

Self: $^9/_{16}$ $+/- ; +/-$

$^3/_{16}$ $+/- ; m^3/m^3$

$^3/_{16}$ $m^1/m^1 ; +/-$

$^1/_{16}$ $m^1/m^1 ; m^3/m^3$

28. These data suggest that three alleles of one gene are being studied and that the order of dominance is black (B) > red (r) > blue (bl).

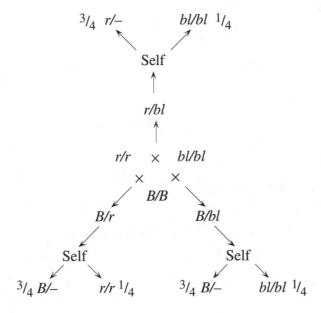

$^3/_4$ $r/-$ bl/bl $^1/_4$

Self

r/bl

$r/r \times bl/bl$

$\times$ $\times$

B/B

B/r B/bl

Self   Self

$^3/_4$ $B/-$ r/r $^1/_4$ $^3/_4$ $B/-$ bl/bl $^1/_4$

29. A single gene showing incomplete dominance (1:2:1 ratio in F_2):

$A/A \times a/a$

$\downarrow$

25% A/A

50% A/a

25% a/a

30. There is a 13:3 ratio of F$_2$ progeny suggesting two genes segregating independently and epistasis.

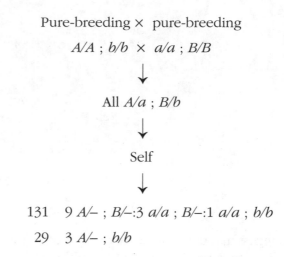

Pure-breeding × pure-breeding

A/A ; b/b × a/a ; B/B

↓

All A/a ; B/b

↓

Self

↓

131   9 $A/-$; $B/-$:3 a/a ; $B/-$:1 a/a ; b/b

29   3 $A/-$; b/b

31. There is a 12:3:1 ratio of F$_2$ progeny suggesting two genes segregating independently and epistasis.

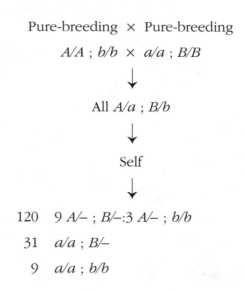

Pure-breeding × Pure-breeding

A/A ; b/b × a/a ; B/B

↓

All A/a ; B/b

↓

Self

↓

120   9 $A/-$; $B/-$:3 $A/-$; b/b

31   a/a ; $B/-$

9   a/a ; b/b

32. These data suggest that three alleles of one gene are being studied. Both codominance (=) and classical dominance (>) are present in this multiple allelic series: l^v (vertical lines) = l^b (horizontal lines), and both l^v and l^b > l^0 (no lines).

$l^v/l^v × l^0/l^0 \rightarrow$ all $l^v/l^0 \rightarrow$ F$_2$ 3 $l^v/-$:1 l^0/l^0

$l^b/l^b × l^0/l^0 \rightarrow$ all $l^b/l^0 \rightarrow$ F$_2$ 3 $l^b/-$:1 l^0/l^0

$l^v/l^v × l^b/l^b \rightarrow$ all $l^v/l^b \rightarrow$ F$_2$ 1 l^v/l^v:2 l^v/l^b:1 l^b/l^b

33. a. The results indicate that two unlinked genes are being studied. One gene determines whether the beetles have a diamond or stripe. For these phenotypes, the diamond allele (D) is dominant to the stripe allele (d). The other gene determines whether the beetle has a spot. In this case, the spot allele (S) is dominant to the no-spot allele (s) and it is also epistatic to D and d.

Cross 1

$$D/d \; ; \; s/s \; \times \; D/d \; ; \; s/s \; \rightarrow \quad {}^1\!/_4 \, D/D \; ; \; s/s + {}^1\!/_2 \, D/d \; ; \; s/s \;\; \text{(diamond)}$$

$${}^1\!/_4 \, d/d \; ; \; s/s \;\; \text{(stripe)}$$

Cross 2

$$D/d \; ; \; S/s \; \times \; d/d \; ; \; S/s \; \rightarrow \quad {}^1\!/_8 \, D/d \; ; \; S/S + {}^1\!/_8 \, d/d \; ; \; S/S + {}^1\!/_4 \, D/d \; ;$$

$$S/s + {}^1\!/_4 \, d/d \; ; \; S/s \;\text{(spot)}$$

$${}^1\!/_8 \, D/d \; ; \; s/s \;\text{(diamond)}$$

$${}^1\!/_8 \, d/d \; ; \; s/s \;\text{(stripe)}$$

Cross 3

$$D/d \; ; \; S/s \; \times \; d/d \; ; \; s/s \; \rightarrow \quad {}^1\!/_4 \, d/d \; ; \; S/s + {}^1\!/_4 \, D/d \; ; \; S/s \;\text{(spot)}$$

$${}^1\!/_4 \, D/d \; ; \; s/s \;\text{(diamond)}$$

$${}^1\!/_4 \, d/d \; ; \; s/s \;\text{(stripe)}$$

b. The spotted progeny represent two genotypes

$${}^1\!/_2 \qquad D/d \; ; \; S/s$$

$${}^1\!/_2 \qquad d/d \; ; \; S/s$$

When intercrossed, there are three possible combinations:

(a) $\quad {}^1\!/_4 \quad D/d \; ; \; S/s \; \times \; D/d \; ; \; S/s$

(b) $\quad {}^1\!/_2 \quad D/d \; ; \; S/s \; \times \; d/d \; ; \; S/s$

(c) $\quad {}^1\!/_4 \quad d/d \; ; \; S/s \; \times \; d/d \; ; \; S/s$

From (a) $\quad {}^3\!/_4$ spot; ${}^1\!/_{16}$ stripe; ${}^3\!/_{16}$ diamond ($\times {}^1\!/_4$ of total crosses)

(b) $\quad {}^3\!/_4$ spot; ${}^1\!/_8$ stripe; ${}^1\!/_8$ diamond ($\times {}^1\!/_2$ of total crosses)

(c) $\quad {}^3\!/_4$ spot; ${}^1\!/_4$ stripe ($\times {}^1\!/_4$ of total crosses)

Weighted total: ${}^3\!/_4$ spot; ${}^9\!/_{64}$ stripe; ${}^7\!/_{64}$ diamond

15 REGULATION OF CELL NUMBER: NORMAL AND CANCER CELLS

1. (1) Certain cancers are inherited as highly penetrant simple Mendelian traits.

   (2) Most carcinogenic agents are also mutagenic.

   (3) Various oncogenes have been isolated from tumor viruses.

   (4) A number of genes that lead to the susceptibility to particular types of cancer have been mapped, isolated, and studied.

   (5) Dominant oncogenes have been isolated from tumor cells.

   (6) Certain cancers are highly correlated to specific chromosomal rearrangements.

2. The activities of proteins controlling the cell cycle or cell-death pathways are controlled by phosphorylation/dephosphorylation, protein-protein interactions, and proteolysis.

3. Cyclins complex with cyclin-dependent protein kinases (CDKs) and activate their kinase activity. They also tether target proteins so that they can be phosphorylated by the complexed CDK.

   The levels of cyclins are regulated by rapid inactivation of the required transcription factors, the high degree of instability of the mRNA encoding the cyclin, and the high degree of instability of the cyclin itself.

4. Oncoproteins may be the result of point mutations, gene fusions, or deletions of key regulatory domains.

5. Apaf is a positive regulator of apoptosis; Bcl is a negative one. Apaf binds to cytoplasmic cytochrome *c* protein, and this complex then binds to and activates the initiator caspase. Bcl blocks the release of cytochrome *c* from the mitochondria and also binds to Apaf, preventing its interaction with the initiator caspase.

6. Translocations can cause the fusion of the enhancer of one gene to the transcription unit of another, leading to the misexpression or overexpression of a nonmutated protein, or they can cause the fusion of two genes resulting in an abnormal chimeric protein.

7. As caspases (cysteine-containing aspartate-specific proteases) are first translated, they contain sequences that prevent their activity. These inactive versions are called zymogens. Part of the polypeptide must then be removed by enzymatic cleavage for the caspase to become active.

8. Translocation and fusion of an immunoglobin enhancer to the *bcl2* gene causes large amounts of Bcl2 to be expressed in B lymphocytes. Since Bcl2 is a negative regulator of apoptosism, this overexpression essentially blocks apoptosis in these mutant B lymphocytes and allows them to accumulate cell proliferation-promoting mutations over their unusually long lifetime.

    Another example would be the point mutation in the gene encoding the Ras protein. The mutant version of this G-protein is always bound to GTP and thus remains active even in the absence of the correct signal.

9. Tumor-promoting mutant alleles of tumor-suppressor genes inactivate the proteins that they encode. Since only when both copies of one of these genes are inactivated are tumors observed, these loss-of-function mutations are by definition recessive. Examples of such genes are the *p53* gene and the *RB* gene.

10. **a.** The v-*erbB* oncogene encodes a mutated form of the epidermal growth factor receptor (EGFR). Compared to wild type, the mutant EGFR oncoprotein lacks key regulatory regions as well as the extracellular ligand-binding domain. These deletions allow the mutant protein to constitutively dimerize, leading to continuous autophosphorylation, which results in continuous transduction of a signal from the receptor.

    b. The *ras* oncogene is due to a missense mutation that allows the mutated Ras protein to always bind GTP, even in the absence of the signals normally required for such binding by wild-type Ras protein. As a result, the Ras oncoprotein continually promotes cell proliferation.

    c. The Philadelphia chromosome is a translocation between chromosomes 9 and 22 that results in the fusion of two genes, *bcr1* and *abl*. The Bcr1-Abl fusion protein has an activated kinase activity that is responsible for causing chronic myelogenous leukemia.

11. **a.** Many growth-factor receptors are receptor tyrosine kinases (RTKs) and are activated by dimerization caused by binding their ligand (the growth factor). Structural aberrations of such a protein can lead to activation in the

absence of ligand or dimerization. In the case of the v-*erbB* oncoprotein, loss of the extracellular ligand-binding domain as well as some regulatory components of the cytoplasmic domain leads to dimerization and continuous activation in the absence of its ligand.

b. Changes in transcriptional regulation that lead to the overproduction or misexpression of an otherwise normal protein may lead to cancer. For example, overproduction of Bcl2 protein (a negative regulator of apoptosis) in B lymphocytes as the result of a translocation that fuses the enhancer of one of the immunoglobulin genes to the transcriptional unit of the *bcl2* gene is the major cause of follicular lymphoma.

c. G proteins cycle between being bound to GDP (inactive state) and being bound to GTP (active state). Mutations that alter the protein preventing its GTPase activity or allowing it to bind to GTP even in the absence of proper signals may cause cancer (as in the case of Ras oncoprotein).

12. a. Cyclins bind to, activate, and direct CDKs to phosphorylate specific cellular targets and by doing so control the cell cycle. Normal cell division requires the sequential production and then removal of different cyclins. The activity of the cyclin-CDK heterodimer is also regulated through p21. Overproduction of one of the cyclins could disrupt the orderly process of cell division, but it would be limited by the amount of CDK present as well as the state of the p21 "brake."

b. A nonsense mutation would lead to a decrease of the normal protein product. If that protein were part of the receptor for a growth factor, which stimulates cell proliferation, then cell division could not be triggered. This would likely be recessive and would slow cell proliferation, not accelerate it.

c. Overproduction of FasL will signal adjacent cells through their Fas cell-surface receptors, which in turn leads to Apaf activation. This in turns causes proteolysis and activation of the initiator caspase, ultimately leading to apoptosis of the cell. Although this would be dominant, it would lead to excess cell death, not proliferation.

d. Cytoplasmic tyrosine-specific protein kinase phosphorylates proteins in response to signals received by the cell. These phosphorylations lead to activation of the transcription factors for the next step in the cell cycle. If the active site is disrupted, then phosphorylations will not occur and transcription factors for the next step will not be activated. This would likely be recessive and would slow cell proliferation, not accelerate it.

e. If the enhancer causes large amounts of the apoptosis inhibitor to be expressed in the liver, the normal pathway of cell death will be blocked. These liver cells (the enhancer is liver-specific) will have an unusually long lifetime in which to accumulate various mutations that could lead to cancer. This chromosomal rearrangement would be dominant.

13. Once apoptosis is initiated, a self-destruct switch has been thrown: endonucleases and proteases are released, DNA is fragmented, and organelles are disrupted.

This obviously is a "terminal" state from which the cell will not have a need or chance to reuse the machinery of destruction. On the other hand, the various proteins needed for the regulation and execution of the cell cycle will be needed again if the cell continues to divide. By recycling many of these, the cell obviously conserves its resources (proteins are energetically expensive to make) and recycling also allows for more rapid divisions, since the cell does not have to spend time remaking all the pieces.

14. Gain-of-function mutations effect mitogenic pathways to increase cell proliferation or cell survival pathways to decrease apoptosis. Loss-of-function mutations promote tumors by loss of growth inhibitor pathways, p53 pathways, or apoptosis activator pathways.

15. **a.** This would be dominant. The misexpression of FasL from one allele would be dominant to the normal expression of the wild-type FasL allele. In this case, each liver cell would signal its neighboring cells to undergo apoptosis.

    b. No. It would lead to excess cell death, not proliferation.

16. Normal Ras is a G-protein that activates a protein kinase, which in turn phosphorylates a transcription factor. If it were simply deleted, no cancer could develop, because cell division would not occur. If it were simply duplicated, an excess of the G-protein could not cause cancer, because it must be activated before it can activate the protein kinase, and presumably the enzyme that activates normal Ras is closely regulated and would not activate too many copies. However, if it were to have a point mutation, it might now bind GTP, even in the absence of normal control signals and be in a state of permanent activation. As a positive regulator of cell growth, this mutant Ras would continually promote cell proliferation.

    In contrast, normal *c-myc* is a transcription factor. If the gene were to be duplicated, too much transcription factor could lead to malignancy.

17. Inhibition of apoptosis can contribute to tumor formation by allowing cells to have an unusually long lifetime in which to accumulate various mutations that lead to cancer. Also, the normal role of apoptosis in removing abnormal cells and, through p53, killing cells that have "damaged" DNA would also be inhibited.

18. **a.** Mutations in a tumor suppressor gene are recessive and due to loss of function. That function can be restored by the introduction of a wild-type allele.

    b. Mutations in an oncogene are dominant and due to gain of function (overexpression or misexpression). The normal function will not inhibit these mutants, and the introduced gene would be ineffective in restoring the normal phenotype.

19. **a.** Type A diabetes is most likely due to a defect in the pancreas. The pancreas normally makes insulin, and type A diabetes can be treated by supplying insulin. Type B diabetes is most likely due to a target cell defect because type B is unresponsive to exogenous insulin.

b. Type B diabetes appears to be caused by a defect in the target cell. A number of genes are responsible for the receptor and the subsequent cascade of changes that occur in leading to a change in transcription. Any of these genes could have a mutant form.

20. p53 detects and is activated by DNA damage. When activated, p53 activates p21, an inhibitor of the cyclin-CDK complex necessary for the progression of the cell cycle. If the DNA damage is repairable, this system will eventually deactivate p53 and allow cell division. However, if the damage is irreparable, p53 would stay active and would activate the apoptosis pathway, ultimately leading to cell death. It is for this reason that the "loss" of p53 is often associated with cancer.

21. Tumors form from cells in which both copies of the *RB* gene are inactivated. Patients with HBR are heterozygous (*RB/rb*) for a mutant copy of the *RB* gene. For these individuals, the mutation rate (and number of retinal cells) makes it virtually certain that at least some of their retinal cells will acquire a mutation in the remaining normal *RB* gene, thereby producing cells with no functional Rb protein. Sporadic retinoblastoma, on the other hand, is the result of a single cell acquiring mutations in both copies of the *RB* gene. Since these mutations are independent events, the occurrence of cells without functional Rb protein will be much rarer.

22. **a.** In the absence of functional Rb protein, E2F will be in the nucleus.

   b. No. The absence of functional E2F would be epistatic to the absence of functional Rb protein.

   c. A mutation that causes permanent sequestering of E2F in the cytoplasm will likely inhibit the cell cycle.

16

THE GENETIC BASIS OF DEVELOPMENT

1. Sex determination in *Drosophila* is autonomous at the cellular level. The *Sxl* gene is permanently turned on or remains off early in development in response to the concentration of X:A transcription factor. Because the X:A ratio is established by the interaction of gene products made in the ovary and in the early zygote, a chemical gradient would be expected to exist that would be sufficiently high in some cells to result in femaleness but low enough in other cells to result in maleness, making the individual an intersex.

2. In humans, a single copy of the Y chromosome is sufficient to shift development toward normal male phenotype. The extra copy of the X chromosome is simply inactivated. Both mechanisms seem to be all-or-none rather than to be based on concentration levels.

3. Because maleness is based on the presence of androgens produced by the developing testes and femaleness is based on the absence of those androgens, what seems to be crucial here is whether the migrating germ cells organize testes. Although what determines this is unknown, it may be that a minimal number of XY cells are required to organize a testis. If, in the mosaic, not enough of these cells exist, then development will be female. If a sufficient number exist, development will be male.

4. The *bcd* mRNA is tethered to the minus ends of microtubules that are located at the anterior pole of the egg. Upon translation, BCD protein diffuses in the common cytoplasm creating the observed gradient.

The *hb-m* mRNA is uniformly distributed throughout the oocyte. However, translation of *hb-m* mRNA is blocked by the NOS protein product. The *nos* mRNA is localized to the posterior pole through its association with the plus ends of microtubules and, upon translation, a posterior to anterior gradient of NOS is generated. This produces the observed anterior to posterior gradient of HB-M protein.

5. CACT and DL form an inactive complex in the cytoplasm. Phosphorylation of the inactive DL and CACT proteins causes conformational changes that break apart the cytoplasmic complex. The now free phosphorylated DL protein is able to migrate into the nucleus where it serves as a transcription factor.

   Rb and E2F proteins are also combined in a cytoplasmic complex that is inactive. Phosphorylation of the Rb protein alters its shape and causes it to dissociate from the E2F protein. The free E2F protein is then able to diffuse into the nucleus and promote transcription.

6. **a.** There must be a diffusible substance produced by the anchor cell that affects development of the six cells. The 1° has the strongest response to the substance, and the 3° represents a lack of response due to a low concentration or absence of the diffusible substance.

   b. Remove the anchor cell and the six equivalent cells. Arrange the six cells in a circle around the anchor cell. All six cells will develop the same phenotype, which will depend on the distance from the anchor cell.

7. **a.** The results suggest that ABa and ABp are not determined at this point in their development. Also, future determination and differentiation of these cells are dependent on their position within the developing organism.

   b. Because an absence of EMS cells leads to a lack of determination and differentiation of AB cells, the EMS cells must be at least in part responsible for AB-cell development, either through direct contact or by the production of a diffusible substance.

   c. Most descendants of the AB cells do not become muscle cells when P2 is present; all descendants of the AB cells become muscle cells when P2 is absent. Therefore, P2 must prevent some AB descendants from becoming muscle cells.

8. Because the receptor is defective, testosterone cannot signal the cell and initiate the cascade of developmental changes that will switch the embryo from the "default" female development to male development. Therefore, the phenotype will be female.

9. Proper *ftz* expression requires *Kr* in the fourth and fifth segments and *kni* in the fifth and sixth segments.

10. The early promoter of the *Sxl* gene is activated by the NUM-NUM transcription factors and is active only early in embryogenesis. Later in embryogenesis and

for the remainder of the life cycle, the *Sxl* gene is transcribed in both sexes from the late promoter. Subject to alternative splicing, the processed transcript from the late promoter encodes active SXL protein only if spliced in the presence of preexisting SXL protein. Therefore, the presence of active SXL protein ensures the further production of more active SXL protein. It is the active SXL product of the early promoter that causes the production of active SXL protein from the late promoter which can then continue to ensure continued production of active SXL protein in the current sex.

11. The anterior-posterior polarity of the *Drosophila* embryo is developed through the action of maternal effect genes. The anterior determinant is the product of the *bcd* gene. *bcd* mRNA is localized to the anterior pole and translation of this localized mRNA creates the anterior-to-posterior gradient of BCD protein. *hb-m* mRNA is uniformly distributed throughout the oocyte. Its translation is blocked by the NOS protein product. *nos* mRNA is localized at the posterior pole and translation of this localized mRNA creates the posterior-to-anterior gradient of NOS protein. The NOS translation repressor posterior-to-anterior gradient produces the shallow anterior-to-posterior gradient of HB-M protein. In embryos lacking functional NOS protein, *hb-m* mRNA will be translated throughout the embryo and posterior segments will be lost. The absence of functional BCD protein allows anterior segments to develop as posterior ones.

12. A number of experiments could be devised. A comparison of amino acid sequence between mammalian gene products and insect gene products would indicate which genes are most similar to each other. Using cloned cDNA sequences from mammalian genes for hybridization to insect DNA would also indicate which genes are most similar to each other.

13. **a.** The anterior 20% of the embryo is normally devoted to the head and thorax regions. The bicaudal phenotype results in the loss of these regions and in the loss of A1 through A3. The gap proteins are responsible for the induction of the pair-rule proteins, which ultimately set the number of segments, and the homeotic proteins, which set the identity of the segments. Obviously, the gap proteins are improperly regulated to produce the bicaudal phenotype.

    Normal regulation of the gap proteins is accomplished by differential sensitivity to the differing concentrations of the maternally derived morphogens. Because the anterior portion of the embryo has been removed, high concentrations of the morphogens in these regions have also been removed. This results in the abnormal segment number and identity that are observed.

    b. The *oskar* mutation results in the loss of the posterior localization of the *nos* mRNA and protein. Therefore, there is no repression of HB-M translation. The lack of repression of the HB-M transcription factor results in an excess of the HB-M protein. The normal shallow gradient, A to P, is therefore lost. Because the gap genes respond differentially to the BCD:HB-M ratio, no induction of the gap genes occurs, which leads to reduced segmentation. This results simply in a broader head and thorax, and no mirror-image phenotype is possible.

14. If you diagram these results, you will see that deletion of a gene that functions posteriorly allows the next-most anterior segments to extend in a posterior direction. Deletion of an anterior gene does not allow extension of the next-most posterior segment in an anterior direction. The gap genes activate *Ubx* in both thoracic and abdominal segments, whereas the *abd-A* and *Abd-B* genes are activated only in the middle and posterior abdominal segments. The functioning of the *abd-A* and *Abd-B* genes in those segments somehow prevents *Ubx* expression. However, if the *abd-A* and *Abd-B* genes are deleted, *Ubx* can be expressed in these regions.

15. It may be that the wild-type allele in the embryo produces a gene product that can inhibit the gene product of the rescuable maternal-effect lethal mutations, while the nonrescuable maternal-effect lethal mutations produce a product that cannot be inhibited.

    Alternatively, the nonrescuable maternal-effect lethal mutations may produce a product that is required very early in development, before the developing fly is producing any proteins, while the rescuable maternal-effect lethal mutations may act later in development when embryo protein production can compensate for the maternal mutation.

16. a. The determination of anterior-posterior portions of the embryo is governed by a concentration gradient of *bcd*. The concentration is highest in the anterior region and lowest in the posterior region. The furrow develops at a critical concentration of *bcd*. As *bcd+* gene dosage (and, therefore, BCD concentration) decreases, the furrow shifts anteriorly; as the gene dosage increases, the furrow shifts posteriorly.

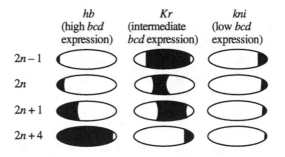

17. Normally, the *tra* gene in the female is active, while in the male it is not active. The active *tra* form of the gene product results in a change in the *dsx* product, shifting development toward the female. If the *dsx* product is not altered, development proceeds along the male line. A mutation in the gene that results in chromosomal females developing as phenotypic, but sterile, males must involve an inactive *tra* product. Homozygotes for the *tra* mutation could be transplanted with male germ cells very early in development, which should result in normal gonad development.

18. The concentration of *Sxl* is crucial for female development and dispensable for male development. The dominant *SxlM* male-lethal mutations may not actually kill all males but simply produce an excessive amount of gene product so that only females (fertile XX and XY) result. The reversions may eliminate all gene product, resulting in XX (sterile) and XY males. The reversions would be recessive because, presumably, a single normal copy of the gene may produce enough gene product to "toggle the switch" in development to female.

19. The anterior-posterior axis would be reversed.

20. a. A pair-rule gene.

   b. Look for expression of the mRNA from the candidate gene in repeating pattern of seven stripes along the A-P axis of developing embryo.

   c. No. An embryo mutant for the gap gene *Krüppel* would be missing many anterior segments and this effect would be epistatic to the expression of a pair-rule gene.

21. There are a number of ways of approaching this problem as various combinations of transgenic strains will give a satisfactory solution. The table below gives all possible combinations and results. Systematically create ten transgenic fruit-fly strains by introducing each of the five plasmids into each of wild type and *bicoid* mutant strains. Observe where the *bicoid* protein is spatially localized in the transgenic embryos. The abbreviations A, P, B, or N are used for anterior, posterior, both, or neither, respectively.

	Transgenic wild type	Transgenic *bicoid* mutant
plasmid 1	A	N
plasmid 2	A	A
plasmid 3	B	P
plasmid 4	A	A
plasmid 5	B	P

22. a. Developmentally, the default sexual pathway in mice is female unless testosterone is able to trigger male development. In the absence of the testosterone receptor, the presence of testosterone cannot be detected by the organism.

   b. Mosaic. Cells that derived from progenitors that had inactivated the wild-type X chromosome would develop along the female pathway, while those derived from progenitors that had inactivated the *Tfm*-carrying X would follow the male pathway.

17

POPULATION GENETICS

1. The frequency of an allele in a population can be altered by natural selection, mutation, migration, nonrandom mating, and genetic drift (sampling errors).

2. There are a total of $(2)(384) + (2)(210) + (2)(260) = 1708$ alleles in the population. Of those, $(2)(384) + 210 = 978$ are *A1* and $210 + (2)(260) = 730$ are *A2*. The frequency of *A1* is $978/1708 = 0.57$, and the frequency of *A2* is $730/1708 = 0.43$.

3. The given data are $q^2 = 0.04$ and $p^2 + 2pq = 0.96$. Assuming Hardy-Weinberg equilibrium, if $q^2 = 0.04$, $q = 0.2$ and $p = 0.8$. The frequency of *B/B* is $p^2 = 0.64$, and the frequency of *B/b* is $2pq = 0.32$.

4. The frequency of a phenotype in a population is a function of the frequency of alleles that lead to that phenotype in the population. To determine dominance and recessiveness, do standard Mendelian crosses.

5. **a.** The needed equations are

$$p' = p\frac{p^W{}_{AA} + q^W{}_{Aa}}{\overline{W}}$$

$$\overline{W} = p^{2W}{}_{AA} + 2pq^W{}_{Aa} + q^{2W}{}_{aa}$$

$$p' = 0.5\,[(0.5)(0.9) + 0.5(1.0)]/[(0.25)(0.9) + (0.5)(1.0) + (0.25)(0.7)] = 0.53$$

b. The needed equation is

$$\hat{p} = \frac{W_{a/a} \times W_{A/a}}{(W_{a/a} - W_{A/a}) + (W_{A/A} - W_{A/a})}$$

$$= \frac{0.7 - 1.0}{(0.7 - 1.0) + (0.9 - 1.0)}$$

$$= 0.75$$

6. The needed equation is

$$q^2 = \mu/s$$

or $\quad s = \mu/q^2 = 10^{-5}/10^{-3} = 0.01$

7. Albinos appear to have had greater opportunity to mate. They may have been considered lucky and encouraged to breed at very high levels in comparison with nonalbinos. They may also have been encouraged to mate with each other. Alternatively, in the tribes with a very low frequency, albinos may have been considered very unlucky and destroyed at birth or prevented from marriage.

8. This problem assumes that there is no backward mutation. Use the following equation:

$$p_n = p_o e^{-n\mu}$$

That is, $p_{50,000} = (0.8)e^{-(5 \times 10^4)(4 \times 10^{-6})} = (0.8)e^{-0.2} = 0.65$

9. **a.** If the variants represent different alleles of gene X, a cross between any two variants should result in a 1:1 progeny ratio (since the organism is haploid). All the variants should map to the same locus. Amino acid sequencing of the variants should reveal differences of one to just a few amino acids.

   b. There could be another gene (gene Y), with five variants, which modifies the gene X product post-transcriptionally. If so, the easiest way to distinguish between the two explanations would be to find another mutation in X and do a dihybrid cross. For example, if there is independent assortment,

   P $\qquad X^1 ; Y^1 \times X^2 ; Y^2$

   $F_1 \qquad 1\ X^1 ; Y^1{:}1\ X^1 ; Y^2{:}1\ X^2 ; Y^1{:}1\ X^2 ; Y^2$

   If the new mutation in X led to no enzyme activity, the ratio would be

   2 no activity:1 variant one activity:1 variant two activity.

   The same mutant in a one-gene situation would yield 1 active:1 inactive.

10. **a.** If the population is in equilibrium, $p^2 + 2pq + q^2 = 1$. Calculate the actual frequencies of p and q in the population and compare their genotypic distribution to the predicted values. For this population:

   $$p = [406 + {}^1/_2(744)]/1482 = 0.52$$

   $$q = [332 + {}^1/_2(744)]/1482 = 0.48$$

The genotypes should be distributed as follows if the population is in equilibrium:

$L^M/L^M = p^2(1482) = 401$ Actual: 406

$L^M/L^N = 2pq(1482) = 740$ Actual: 744

$L^N/L^N = q^2(1482) = 341$ Actual: 332

This compares well with the actual data, so the population is in equilibrium.

b. If mating is random with respect to blood type, then the following frequency of matings should occur:

$L^M/L^M \times L^M/L^M = (p^2)(p^2)(741) = 54$ Actual: 58

$L^M/L^M \times L^M/L^N$ or $L^M/L^N \times L^M/L^M = (2)(p^2)(2pq)(741) = 200$ Actual: 202

$L^M/L^N \times L^M/L^N = (2pq)(2pq)(741) = 185$ Actual: 190

$L^M/L^M \times L^N/L^N$ or $L^N/L^N \times L^M/L^M = (2)(p^2)(q^2)(741) = 92$ Actual: 88

$L^M/L^N \times L^N/L^N$ or $L^N/L^N \times L^M/L^N = 2(2pq)(q^2)(741) = 170$ Actual: 162

$L^N/L^N \times L^N/L^N = (q^2)(q^2)(741) = 39$ Actual: 41

Again, this compares nicely with the actual data, so the mating is random with respect to blood type.

11. a., b. For each, p and q must be calculated and then compared with the predicted genotypic frequencies of $p^2 + 2pq + q^2 = 1$.

Population	p	q	Equilibrium?
1	1.0	0.0	Yes
2	0.5	0.5	No
3	0.0	1.0	Yes
4	0.625	0.375	No
5	0.375	0.625	No
6	0.5	0.5	Yes
7	0.5	0.5	No
8	0.2	0.8	Yes
9	0.8	0.2	Yes
10	0.993	0.007	Yes

c. The formulas to use are $q^2 = \mu/s$ and $s = 1 - W$.

$$4.9 \times 10^{-5} = 5 \times 10^{-6}/s\,;\ \ s = 0.102,\ \text{so}\ W = 0.898$$

d. For simplicity, assume that the differences in survivorship occur prior to reproduction. Thus, each genotype's fitness can be used to determine the relative percentage each contributes to the next generation.

Genotype	Frequency	Fitness	Contribution	A	a
A/A	0.25	1.0	0.25	0.25	0.0
A/a	0.50	0.8	0.40	0.20	0.20
a/a	0.25	0.6	0.15	0.0	0.15
			SUM:	0.45	0.35

$$p' = 0.45/(0.45 + 0.35) = 0.56$$

$$q' = 0.35/(0.45 + 0.35) = 0.44$$

Alternatively, the formulas to use are

$$p' = p\frac{pW_{AA} + qW_{Aa}}{\overline{W}}$$

$$\overline{W} = p2W_{AA} + 2pqW_{Aa} + q^2W_{aa}$$

$$p' = (0.5)[(0.5)(1.0) + (0.5)(0.8)]/[(0.25)(1.0) + (0.5)(0.8) + (0.25)(0.6)]$$

$$= (0.5)(0.9)/(0.8) = 0.56$$

12. **a.** Assuming the population is in Hardy-Weinberg equilibrium and that the allelic frequency is the same in both sexes, we can directly calculate the frequency of the color-blind allele as $q = 0.1$. (Since this trait is sex-linked, q is equal to the frequency of affected males.) Color-blind females must be homozygous for this X-linked recessive trait, so their frequency in the population is equal to $q^2 = 0.01$.

b. There would be 10 color-blind men for every color-blind woman $(^q/_q2)$.

c. For this condition to be true, the mothers must be heterozygous for the trait and the fathers must be color-blind ($X^C/X^c \times X^c/Y$). The frequency of heterozygous women in the population will be $2pq$, and the frequency of color-blind men will be q. Therefore, the frequency of such random marriages will be $(2pq)(q) = 0.018$.

d. All children will be phenotypically normal only if the mother is homozygous for the noncolor-blind allele ($p^2 = 0.81$). The father's genotype does not matter and therefore can be ignored.

e. There are several ways of approaching this problem. One way to visualize the data, however, is to construct the following chart:

Father Mother	0.4 X^C	0.6 X^c	Y
0.8 X^C	0.32 X^C/X^C	0.48 X^C/X^c	0.8 X^C/Y
0.2 X^c	0.08 X^C/X^c	0.12 X^c/X^c	0.2 X^c/Y

As can be seen, the frequency of color-blind females will be 0.12 and color-blind males 0.2.

f. From analysis of the results in part e, the frequency of the color-blind allele will be 0.2 in males (the same as in the females of the previous generation) and $\frac{1}{2}(0.08 + 0.48) + 0.12 = 0.4$ in females.

13. Assume that proper function results from the right gene products in the proper ratio to all other gene products. A mutation will change the gene product, eliminate the gene product, or change the ratio of it to all other gene products. All three outcomes upset a previously balanced system. While a new and "better" balance may be achieved, this is less likely than being deleterious.

14. Wild-type alleles are usually dominant because most mutations result in lowered or eliminated function. To be dominant, the heterozygote has approximately the same phenotype as the dominant homozygote. This will typically be true when the wild-type allele produces a product and the mutant allele does not.

   Chromosomal rearrangements are often dominant mutations because they can cause gross changes in gene regulation or even cause fusions of several gene products. Novel activities, overproduction of gene products, etc., are typical of dominant mutations.

15. Prior to migration, $q^A = 0.1$ and $q^B = 0.3$ in the two populations. Since the two populations are equal in number, immediately after migration, $q^{A+B} = \frac{1}{2}(q^A + q^B) = \frac{1}{2}(0.1 + 0.3) = 0.2$. At the new equilibrium, the frequency of affected males is $q = 0.2$, and the frequency of affected females is $q^2 = (0.2)^2 = 0.04$. (Color-blindness is an X-linked trait.)

16. For a population in equilibrium, the probability of individuals being homozygous for a recessive allele is q^2. Thus, for small values of q, few individuals in a randomly mating population will express the trait. However, if two individuals share a close common ancestor, there is an increased chance of homozygosity by descent, since only one "progenitor" need be heterozygous.

   For the following, it is assumed that the allele in question is rare. Thus, the chance of both "progenitors" being heterozygous will be ignored.

   a. For a parent-offspring mating, the pedigree can be represented as follows:

   In this example, it is only the chance of the incestuous parent's being heterozygous that matters. Thus, the chance of the descendants being homozygous is

$$2pq(\tfrac{1}{2})(\tfrac{1}{4}) = \tfrac{pq}{4}$$

If q is very small, then p is nearly 1.0 and the chance of an affected child can be represented as approximately $q/4$. (Again, this should be compared to the expected random-mating frequency of q^2.)

b. For a mating of first cousins, the pedigree can be represented as follows:

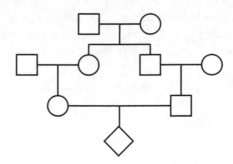

The probability of inheriting the recessive allele if *either* grandparent is heterozygous can be represented as follows:

$$2pq \quad \text{or} \quad 2pq$$

$$\begin{array}{c} \tfrac{1}{2} \quad \tfrac{1}{2} \\ \tfrac{1}{2} \quad \quad \tfrac{1}{2} \\ \tfrac{1}{2} \quad \quad \tfrac{1}{2} \\ \tfrac{1}{4} \end{array}$$

Thus, the chance of this child being affected is

$$2pq \, (^1/_2)(^1/_2)(^1/_2)(^1/_2)(^1/_4) + 2pq \, (^1/_2)(^1/_2)(^1/_2)(^1/_2)(^1/_4) = {}^{pq}/_{16}$$

Again, if q is rare, p is nearly 1.0, so the chance of homozygosity by descent is approximately $q/16$.

c. An aunt-nephew (or uncle-niece) mating can be represented as:

Following the possible inheritance of the recessive allele from either grandparent,

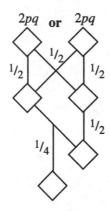

The chance of this child being homozygous is

$$2pq(^1/_2)(^1/_2)(^1/_2)(^1/_4) + 2pq(^1/_2)(^1/_2)(^1/_2)(^1/_4) = {}^{pq}/_8,\text{ or for rare alleles}$$
approximately $^q/_8$.

17. a. The allele frequencies are $f(A) = 0.2 + {}^1/_2(0.60) = 50\%$

$$f(a) = {}^1/_2(0.60) + 0.2 = 50\%$$

b. *Positive assortative mating*: the alleles will randomly unite within the same phenotype. For $A/{-}$, the mating population is $0.2\ A/A + 0.6\ A/a$. The allelic frequencies within this subpopulation are

$$f(A) = [0.2 + {}^1/_2(0.6)]/0.8 = 0.625$$

$$f(a) = {}^1/_2(0.6)/0.8 = 0.375$$

The phenotypic frequencies that result are

$A/{-}$: $p^2 + 2pq = (0.625)^2 + 2(0.625)(0.375) = 0.39 + 0.47 = 0.86$

a/a: $q^2 = (0.375)^2 = 0.14$

However, assuming that all contribute equally to the next generation and this subpopulation represents 0.8 of the total population, these figures must be adjusted to reflect this weighting:

$A/{-}$: $(0.86)(0.8) = 0.69$

a/a: $(0.14)(0.8) = 0.11$

The a/a contribution from the other subpopulation will remain unchanged because there is only one genotype, a/a. Its weighted contribution to the total phenotypic frequency is 0.20. Therefore, after one generation, the phenotypic frequencies will be $A/{-} = 0.69$ and $a/a = 0.20 + 0.11 = 0.31$, and the genotypic frequencies will be $f(A) = 0.5$ and $f(a) = 0.5$. Over time, these allelic frequencies will stay the same, but the frequency of heterozygotes will continue to decrease until there are two separate populations, A/A and a/a, which will not interbreed.

    c. *Negative assortative mating*: mating is between unlike phenotypes. The two types of progeny will be *A/a* and *a/a*. *A/A* will not exist. *A/a* will result from all *A/A* × *a/a* matings and half the *A/a* × *a/a* matings. These matings will occur with the following relative frequencies:

$$A/A \times a/a = (0.2)(0.2) = 0.04$$

$$A/a \times a/a = (0.6)(0.2) = 0.12$$

Because these are the only matings that will occur, they must be put on a 100% basis by dividing by the total frequency of matings that occur:

    A/A × *a/a*:   0.04/0.16 = 0.25, all of which will be *A/a*

    A/a × *a/a*:   0.12/0.16 = 0.75, half *A/a* and half *a/a*

The phenotypic frequencies in this generation will be

    A/a: 0.25 + 0.75/2 = 0.625

    a/a: 0.75/2 = 0.375

    d. In the next generation, since all matings are now between heterozygotes and homozygous recessives, the final allelic frequencies of $f(A) = 0.25$ and $f(a) = 0.75$ will be obtained and the population will be 50% *A/a* and 50% *a/a*.

18. **a.** Many genes affect bristle number in *Drosophila*. The artificial selection resulted in lines with mostly high-bristle-number alleles. Some mutations may have occurred during the 20 generations of selective breeding, but most of the response was caused by alleles present in the original population. Assortment and recombination generated lines with more high-bristle-number alleles.

    b. Fixation of some alleles causing high bristle number would prevent complete reversal. Some high-bristle-number alleles would have no negative effects on fitness, so there would be no force pushing bristle number back down because of those loci.

    c. The low fertility in the high-bristle-number line could have been a result of pleiotropy or linkage. Some alleles that caused high bristle number may also have caused low fertility (pleiotropy). Chromosomes with high-bristle-number alleles may also carry alleles at different loci that caused low fertility (linkage). After artificial selection was relaxed, the low-fertility alleles would have been selected against through natural selection. A few generations of relaxed selection would have allowed low-fertility-linked alleles to recombine away, producing high-bristle-number chromosomes that did not contain low-fertility alleles. When selection was reapplied, the low-fertility alleles had been reduced in frequency or separated from the high-bristle loci, so this time there was much less of a fertility problem.

19. Affected individuals $= B/b = 2pq = 4 \times 10^{-6}$. Because q is almost equal to 1.0, $2p = 4 \times 10^{-6}$. Therefore, $p = 2 \times 10^{-6}$.

$$\mu = hsp = (1.0)(0.7)(2 \times 10^{-6}) = 1.4 \times 10^{-6}$$

where h = degree of dominance of the deleterious allele.

20. The probability of not getting a recessive lethal genotype for one gene is $1 - \frac{1}{8}$ $= \frac{7}{8}$. If there are n lethal genes, the probability of not being homozygous for any of them is $(\frac{7}{8})^n = \frac{13}{31}$. Solving for n, an average of 6.5 recessive lethals is predicted.

   If the actual percentage of "normal" children is less owing to missed in utero fatalities, the average number of recessive lethals would be higher.

21. a. The formula needed is

$$\hat{q} = \sqrt{\mu/s}$$

$$= 4.47 \times 10^{-3}$$

Genetic cost $= sq^2 = 0.5(4.47 \times 10^{-3})^2 = 10^{-5}$

b. Using the same formulas as in part a,

$$\hat{q} = 6.32 \times 10^{-3}$$

Genetic cost $= sq^2 = 0.5(6.32 \times 10^{-3})^2 = 2 \times 10^{-5}$

c.

$$\hat{q} = 5.77 \times 10^{-3}$$

Genetic cost $= sq^2 = 0.3(5.77 \times 10^{-3})^2 = 10^{-5}$

18
QUANTITATIVE GENETICS

1. There are many traits that vary more or less continuously over a wide range. For example, height, weight, shape, color, reproductive rate, metabolic activity, etc., vary quantitatively rather than qualitatively. Continuous variation can often be represented by a bell-shaped curve, where the "average" phenotype is more common than the extremes. Discontinuous variation describes the easily classifiable, discrete phenotypes of simple Mendelian genetics: seed shape, auxotrophic mutants, sickle-cell anemia, etc. These traits show a simple relationship between genotype and phenotype.

2. The mean (or average) is calculated by dividing the sum of all measurements by the total number of measurements, or in this case, the total number of bristles divided by the number of individuals.

   Mean = $\bar{x}$ = [1 + 4(2) + 7(3) + 31(4) + 56(5) + 17(6) + 4(7)]/(1 + 4 + 7 + 31 + 56 + 17 + 4)

   $= {}^{564}/_{120}$ = 4.7 average number of bristle/individual

   The variance is useful for studying the distribution of measurements around the mean and is defined in this example as

   Variance = s^2 = average of the (actual bristle count – mean)^2

   $s^2 = {}^1/_N \Sigma(x_i - x)^2$

   $= {}^1/_{120} \Sigma[(1 - 4.7)^2 + (2 - 4.7)^2 + (3 - 4.7)^2 + (4 - 4.7)^2 + (5 - 4.7)^2 + (6 - 4.7)^2 + (7 - 4.7)^2]$

   $= 0.26$

The standard deviation, another measurement of the distribution, is simply calculated as the square root of the variance:

$$\text{Standard deviation} = s = \sqrt{0.26} = 0.51$$

3. **a.** H^2 has meaning only with respect to the population that was studied in the environment in which it was studied. Even if a trait shows high heritability, it does not imply the trait is unaffected by its environment. The only acceptable analysis is to study directly the norms of reaction of the various genotypes in the population over the range of projected environments. Since it is so difficult to fully replicate a human genotype so that it might be tested in different environments, there is no known norm of reaction for any human quantitative trait.

   b. Neither H^2 nor b^2 is a reliable measure that can be used to generalize from a particular sample to a "universe" of the human population. They certainly should not be used in social decision making (as implied by the terms eugenics and dysgenics).

   c. Again, H^2 and b^2 are not reliable measures, and they should not be used in any decision making with regard to social problems.

4. The following are unknown: (1) norms of reaction for the genotypes affecting IQ, (2) the environmental distribution in which the individuals developed, and (3) the genotypic distributions in the populations. Even if the above were known, because heritability is specific to a specific population and its environment, the difference between two different populations cannot be given a value of heritability.

5. **a.** Broad heritability measures that portion of the total variance that is a result of genetic variance. The equation to use is

   $$H^2 = \text{the genetic variance/phenotypic variance}$$

   where genetic variance = phenotypic variance – environmental variance

   $$H^2 = \frac{s_p^2 - s_e^2}{s_p^2}$$

   Narrow heritability measures that portion of the total variance that is a result of the additive genetic variation. The equation to use is

   $$b^2 = \frac{\text{additive genetic variance}}{\text{additive genetic variance + dominance variance + environmental variance}}$$

   $$b^2 = \frac{s_a^2}{s_a^2 + s_d^2 + s_e^2}$$

Shank length:

$$H^2 = (310.2 - 248.1)/(310.2) = 0.200$$

$$h^2 = 46.5/(46.5 + 15.6 + 248.1) = 0.150$$

Neck length:

$$H^2 = (730.4 - 292.2)/(730.4) = 0.600$$

$$h^2 = 73.0/(73.0 + 365.2 + 292.2) = 0.010$$

Fat content:

$$H^2 = (106.0 - 53.0)/106.0 = 0.500$$

$$h^2 = 42.4/(42.4 + 10.6 + 53.0) = 0.400$$

b. The larger the value of h^2, the greater the difference between selected parents and the population as a whole and the more that characteristic will respond to selection. Therefore, fat content would respond best to selection.

c. The formula needed is

$$\text{Selection response} = h^2 \times \text{selection differential}$$

Therefore, selection response = $(0.400)(10.5\% - 6.5\%) = 1.6\%$ decrease in fat content, or 8.9% fat content.

6. a. The probability of any gene being homozygous is $1/2$ (e.g., for A: A/A or a/a), and the probability of being heterozygous (or not homozygous) is also $1/2$. Thus, the probability for any one gene being homozygous while the other two are heterozygous is $(1/2)^3$. Since there are three ways for this to happen (homozygosity at A or at B or at C), the total probability is

$$p(\text{homozygous at 1 locus}) = 3(1/2)^3 = 3/8$$

The same logic can be applied to any two genes being homozygous:

$$p(\text{homozygous at 2 loci}) = 3(1/2)^3 = 3/8$$

There are two ways for all three genes to be homozygous, so

$$p(\text{homozygous at 3 loci}) = 2(1/2)^3 = 2/8$$

b.

$$p(0 \text{ capital letters}) = p(\text{all homozygous recessive}) = (1/4)^3 = 1/64$$

$$p(1 \text{ capital letter}) = p(1 \text{ heterozygote and 2 homozygous recessive}) =$$
$$3(1/2)(1/4)(1/4) = 3/32$$

$$p(2 \text{ capital letters}) = p(1 \text{ homozygous dominant and 2 homozygous recessive})$$

or

$$p(2 \text{ heterozygotes and 1 homozygous recessive})$$

$$= 3(1/4)^3 + 3(1/4)(1/2)^2 = 15/64$$

$$p(3 \text{ capital letters}) = p(\text{all heterozygous})$$

or

$p(1 \text{ homozygous dominant, 1 heterozygous, and}$
$1 \text{ homozygous recessive})$

$$= (^1/_2)^3 + 6(^1/_4)(^1/_2)(^1/_4) = {}^{10}/_{32}$$

$p(4 \text{ capital letters}) = p(2 \text{ homozygous dominant and 1 homozygous recessive})$

or

$p(1 \text{ homozygous dominant and 2 heterozygous})$

$$= 3(^1/_4)^3 + 3(^1/_4)(^1/_2)^2 = {}^{15}/_{64}$$

$p(5 \text{ capital letters}) = p(2 \text{ homozygous dominant and 1 heterozygote})$

$$= 3(^1/_4)^2(^1/_2) = {}^3/_{32}$$

$p(6 \text{ capital letters}) = p(\text{all homozygous dominant}) = (^1/_4)^3 = {}^1/_{64}$

7. For three genes there are a total of 27 genotypes that will occur in predictable proportions. For example, there are 3 genotypes that have two genes that are heterozygous and one gene that is homozygous recessive (*A/a* ; *B/b* ; *c/c, A/a* ; *b/b* ; *C/c, a/a* ; *B/b* ; *C/c*). The frequency of this combination is $3(^1/_2)(^1/_2)(^1/_4)$ $= {}^3/_{16}$, and the phenotypic score is $3 + 3 + 1 = 7$. For all the genotypes possible, the total distribution of phenotypic scores is as follows:

Score	Proportion
3	$^1/_{64}$
5	$^3/_{32}$
6	$^3/_{64}$
7	$^3/_{16}$
8	$^3/_{16}$
9	$^{11}/_{64}$
10	$^3/_{16}$
11	$^3/_{32}$
12	$^1/_{64}$

And the plot of these data will be

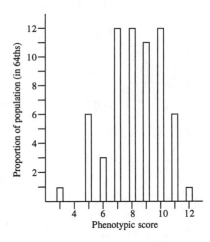

8. The population described would be distributed as follows:

3 bristles $^{19}/_{64}$

2 bristles $^{44}/_{64}$

1 bristle $^{1}/_{64}$

The 3-bristle class would contain 7 different genotypes, the 2-bristle class would contain 19 different genotypes, and the 1-bristle class would contain only 1 genotype. It would be very difficult to determine the underlying genetic situation by doing controlled crosses and determining progeny frequencies.

9. **a.** Solving the formula for values of x over the stated range for each genotype gives the following data:

x	1	2	3
0.03	0.90		
0.04	0.91		
0.05	0.93		
0.06	0.93		
0.07	0.94		
0.08	0.95		
0.09	0.96		
0.10	0.97		0.90
0.11	0.97		0.92
0.12	0.98	0.90	0.93

x	1	2	3
0.13	0.98	0.92	0.94
0.14	0.99	0.94	0.95
0.15	0.99	0.95	0.96
0.16	0.99	0.96	0.97
0.17	1.00	0.98	0.98
0.18	1.00	0.98	0.98
0.19	1.00	0.99	0.99
0.20	1.00	1.00	0.99
0.21	1.00	1.00	1.00
0.22	1.00	1.00	1.00
0.23	1.00	1.00	1.00
0.24	0.99	1.00	1.00
0.25	0.99		1.00
0.26	0.99		1.00
0.27	0.98		1.00
0.28	0.98		0.99
0.29	0.97		0.99
0.30	0.97		0.98
0.31	0.96		0.98
0.32	0.95		0.97
0.33	0.94		0.96
0.34	0.93		0.95
0.35	0.93		0.94
0.36	0.91		0.93
0.37	0.90		0.92
0.38			0.90

Plotting these data give the following curves:

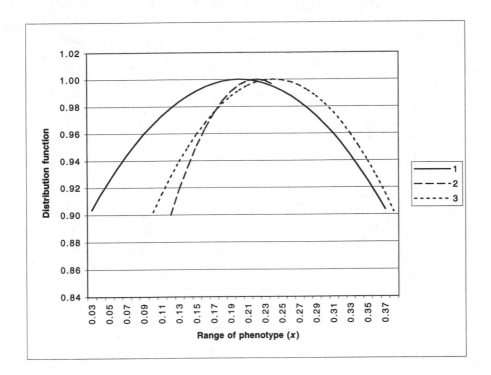

b. Since the 3 genotypes are equally frequent, the average distribution across the entire range of phenotypes will be

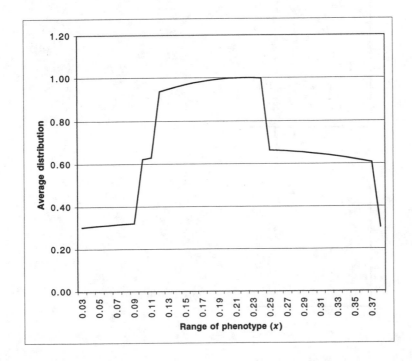

There are regions of the overall phenotypic distribution where the variation within a given genotype does not overlap, and this gives sharp steps to the distribution. On the other hand, any individual whose phenotype lies between values of 0.12 to 0.24 could have any of the 3 genotypes.

10. a.

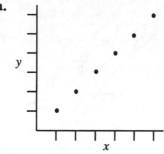

b.

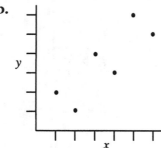

c.

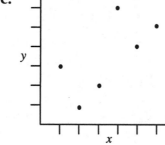

d.

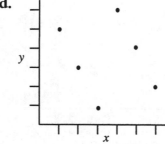

Use the following formulas to calculate the correlation coefficient (r_{xy}) between x and y:

$$r_{xy} = \frac{\text{cov } xy}{s_x s_y} \quad \text{and cov } xy = \frac{1}{N}\Sigma x_i y_i - \overline{xy}$$

a. cov $xy = \frac{1}{6}[(1)(1) + (2)(2) + (3)(3) + (4)(4) + (5)(5) + (6)(6)] - (\frac{21}{6})(\frac{21}{6})$
   $= 2.92$

Standard deviation $x = s_x = \sqrt{\frac{1}{N}\Sigma(x_i - \overline{x})^2} = 1.71$

Standard deviation $y = s_y = \sqrt{\frac{1}{N}\Sigma(y_i - \overline{y})^2} = 1.71$

Therefore, $r_{xy} = 2.92/(1.71)(1.71) = 1.0$. The other correlation coefficients are calculated in a like manner.

b. 0.83

c. 0.66

d. −0.20

11. First, define alcoholism in behavioral terms. Next, realize that all observations must be limited to the behavior you used in the definition and all conclusions from your observations are applicable only to that behavior. In order to do your data gathering, you must work with a population in which familiality is distinguished from heritability. In practical terms, this means using individuals who are genetically close but who are found in all environments possible.

12. Before beginning, it is necessary to understand the data. The first entry, h/h h/h, refers to the II and III chromosomes, respectively. Thus, there are four h (high bristle number) sets of alleles in two or more genes on separate chromosomes. The next entry is h/l h/h. Chromosome II is now heterozygous and chromosome III is still homozygous, etc.

  The effect of substituting one l chromosome II for an h chromosome II, and therefore going from homozygous h/h to heterozygous h/l, can be seen in the differences along the rows in the first two columns. The average change is (2.9 + 3.1 + 2.7)/3 = 2.9. When chromosome II goes from heterozygous h/l to homozygous l/l, the average change is (3.2 + 5.2 + 6.8)/3 = 5.1.

  The effect of substituting one l chromosome III for an h chromosome III can be seen in the differences between rows: 25.1 − 23.0 = 2.1; 22.2 − 19.9 = 2.3; and 19.0 − 14.7 = 4.3 (average change 2.9). And going from h/l to l/l for chromosome III gives an average change of (11.2 + 10.8 + 12.4)/3 = 11.5 bristles.

  Here is a summary of these results:

	Chromosome II	Chromosome III
h/h to h/l	2.9	2.9
h/l to l/l	5.1	11.5

Each set of alleles for both chromosomes is expressed in the phenotype, but that expression varies with the chromosome. Chromosome III appears to have a stronger effect on the phenotype than does chromosome II. (Compare h/h h/h with both l/l h/h and h/h l/l. The difference in the first case is 6.1, and in the second case, 13.3.) Finally, there is partial dominance of h over l for both chromosomes. The change from h/h to h/l is less than the change from h/l to l/l.

13. **a.**

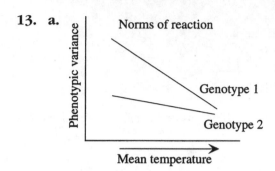

b. Broad heritability is defined as

$$H^2 = \frac{s_g^2}{s_P^2}$$

Assuming that the genetic variance stays the same, phenotypic variance must decrease if H^2 increases. Therefore, the same plot as in part a will satisfy the conditions.

c. To satisfy the conditions that genetic variance increases as H^2 decreases, phenotypic variance must also increase. Therefore, the plot will be

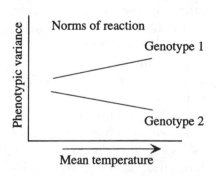

d.

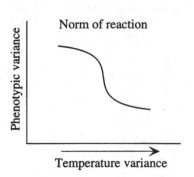

14. **a.** The regression line shows the relationship between the two variables. It attempts to predict one (the son's height) from the other (the father's height). If the relationship is perfectly correlated, the slope of the regression line should approximate 1. If you assume that individuals at the extreme of any spectrum are homozygous for the genes responsible for these phenotypes, then their offspring are more likely to be heterozygous than are the original individuals. That is, they will be less extreme. Also, there is no attempt to include the maternal contribution to this phenotype.

   b. For Galton's data, regression is an estimate of heritability (h^2), *assuming* that there were few environmental differences between all fathers and all sons both individually and as a group. However, there is no evidence given to determine if the traits are familial but not heritable. This data would indicate genetic variation only if the relatives do not share common environments more than nonrelatives do.

19

Evolutionary Genetics

1. A transformational scheme of evolution predicts that all members of a species will change over time. By sharing similar environments and life experiences, each member is altered in its lifetime and progeny inherit these alterations. For example, if all giraffes stretch their necks during their lifetimes to reach the ever-higher foliage, all offspring will inherit this acquired "stretched" neck and begin their lives where their parents have left off. Over time, longer and longer-necked giraffes would evolve.

   A variational scheme of evolution predicts that not all members of a species contribute equally to future generations. Heritable differences among members of the same species result in some being more "fit" and able to produce more offspring than others. Over time, one type of individual is replaced by another. For example, using an economic metaphor, there is currently an explosion of internet and "dot.com" companies. A variation scheme of evolution would suggest that some of these companies will grow, evolve, and prosper while others, based on intrinsic (heritable) differences (management, capitalization, etc.) will become extinct.

2. The three principles are: (1) organisms within a species vary from one another, (2) the variation is heritable, and (3) different types leave different numbers of offspring in future generations.

3. The variation required by Darwin's proposed mechanism of evolution can not exist or be maintained if "blending" or nonrandom segregation occurs. In either case, populations will become homogeneous and variation would be rapidly

lost. The "particulate" nature of genes described by Mendel allows for segregation of traits generation after generation. In this way, even currently detrimental (and recessive) alleles can be retested as the environment and circumstances change.

4. A geographical race is a population that is genetically distinguishable from other local populations but is capable of exchanging genes with those other local populations.

    A species is a group of organisms that exchanges genes within the group but cannot do so with other groups.

    Populations that are geographically separated will diverge from each other as a consequence of a combination of unique mutations, selection, and genetic drift. For populations to diverge enough to become reproductively isolated, spatial separation sufficient to prevent any effective migration is usually necessary.

5. A population will not differentiate from other populations by local inbreeding if

$$\mu \geq 1/N$$

so $$N \geq 1/\mu$$

$$N \geq 105$$

6. The rate of loss of heterozygosity in a closed population is $1/(2N)$ per generation.

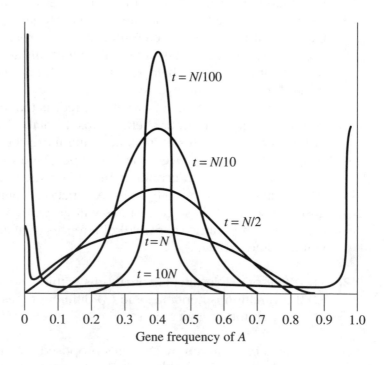

7. A population will not differentiate from other populations by local inbreeding if the number of migrant individuals ≥ 1 per generation.

   For (a), migration is not sufficient to prevent local inbreeding so the results are roughly the same as seen in problem 5. In the case of (b), there is one migrant per generation so the populations will not differentiate and allelic frequencies will remain the same in all populations.

8. The mean fitness is calculated by summing the frequency of each progeny class times its fitness. For example, the frequency of $A/A \cdot B/B$ is $p(A)2 \times p(B)2$ or $(0.64)(0.81) = 0.52$. This frequency is multiplied by its fitness to give $(0.52 \times 0.95) = 0.49$. Summing for all classes, the mean fitness of population 1 [$p(A) = 0.8$, $p(B) = 0.9$] is 0.92. Since selection acts to increase the mean fitness, the frequency of both A and B should increase in the next generation (the $A/A \cdot BB$ class has a fitness of 0.95).

   The mean fitness of population 2 [$p(A) = 0.2$, $p(B) = 0.2$] is 0.73. Again, the frequency of both A and B should increase.

   There is a single adaptive peak at $A/A \cdot B/B$. By inspection, the fitness is lowest at $a/a \cdot b/b$ and highest at $A/A \cdot B/B$. The allelic frequencies at the peak is 1.0 for both A and B.

9. The mean fitness for population 1 [$p(A) = 0.5$, $p(B) = 0.5$] is 0.825. The mean fitness for population 2 [$p(A) = 0.1$, $p(B) = 0.1$] is 0.856. There are four adaptive peaks: at $p(A) = 0.0$ or 1.0 and $p(B) = 0.0$ or 1.0. (The mean fitness will be 0.90 at any of these points.) With population 1, the direction of change for both $p(A)$ and $p(B)$ will be random. Both higher or lower frequencies of either allele can result in increased mean fitness (although there are some combinations that would lower fitness). Since population 2 is already near an adaptive peak of $p(A) = 0.0$ and $p(B) = 0.0$, both $p(A)$ and $p(B)$ should decrease to increase the mean fitness.

10. Above a chromosome number of 12, even numbers are much more common than odd numbers. This is evidence of frequent polyploidization during plant evolution. For example, if a species of plant with an odd haploid chromosome number undergoes a "doubling" event, the chromosome number becomes even.

11. The α and β gene families show remarkable amino acid sequence similarities (see Table 19-4 of the text). Within each gene family, sequence similarities are greater and in some cases, member genes have identical intron-exon structure.

12. All human populations have high i, intermediate I^A, and low I^B frequencies. The variations that do exist among the different geographical populations are most likely due to genetic drift. There is no evidence that selection plays any role regarding these alleles.

13. To test the species distinctness, it is necessary to be able to manipulate and culture *D. pseudoobscura* and *D. persimilis* in captivity. If they cannot be cultured in the laboratory their species distinctness cannot be established. The mating behavior compatibility of the different *Drosophila* can be tested by placing a mixture of males of both forms with females of one of the forms to see whether there are any female mating preferences. The same experiment can then be repeated with mixed females and one sort of male and with mixtures of males and females of both forms. From such experiments, patterns of mating preference can be observed. Even if there is some small amount of mating of different forms, this may occur only because of the unnatural conditions in which the test is being carried out. On the other hand, no mating of any kind may occur, even between the same forms, because the necessary cues for mating are missing, in which case nothing can be concluded.

If matings between different forms do occur, the survivorship of the interpopulation hybrids can be compared with that of the intrapopulation matings. If hybrids survive, their fertility can be tested by attempting to back-cross them to the two different parental strains. As in the case of the mating tests, under the unnatural conditions of the laboratory, some survivorship or fertility of species hybrids is possible even though the isolation in nature is complete. Any clear reduction in observed survivorship or fertility of the hybrids is strong presumptive evidence that they belong to different species.

14.

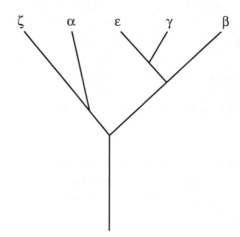

15. For polymorphic sites within a species, let nonsynonymous = *a* and synonymous = *b*. For polymorphic sites between the species, let nonsynonymous = *c* and synonymous = *d*. If divergence is due to neutral evolution, then

$$a/b = c/d$$

If divergence is due to selection, then

$$a/b < c/d$$

However, in this example, $a/b = 20/50 > c/d = 2/18$, which fits neither expectation.

Since the ratio of nonsynonymous to synonymous polymorphisms (*a/b*) is relatively high, the gene being studied may encode a protein tolerant of substitution (like fibrinopeptides, see figure 26-18 of the text). The relatively fewer species differences may suggest that speciation was a recent event so few polymorphisms have been fixed in one species that are not variants in the other.

16. Morphological, anatomical, and developmental studies would help indicate whether the tails derived from the same basic structures. Also, comparison of the genes and proteins involved in the morphogenesis of the tails will determine if common elements underlie their development.

INTERACTIVE GENETICS

Detailed Table of Contents

Mendelian Analysis

4- Probabilities and pedigrees (PTC)
- Probability offspring having dominant trait or recessive trait
- Combination of inheriting trait for particular sex
- Probability of multiple children unaffected

5- Probabilities and pedigrees (rare rec. trait)
- Probability affected offspring
- Dizygotic twins versus monozygotic twins

6- Probabilities and pedigrees (PKU)
- Probability affected offspring
- Probability two out of next five children affected
- Probability of no more than one affected child out of five

7- Ascertainment bias (albinism)
- Real data versus predicted ratios

Chromosome Theory of Inheritance

Tutorials (on CD-ROM)
Mitosis
- Animation and description of phases of mitosis

Meiosis
- Animation and description of phases of meiosis

Mitosis and Meiosis Problems
1- Mitosis
- Stages of cell cycle
- "c" versus "n"
- DNA replication, sister chromatids, homologous chromosomes
- Mitotic division phases
- Alleles on sister chromatids
- Drag and drop chromosomes to emulate mitosis

2- Meiosis
- Consequences of sexual reproduction without meiosis
- Meiosis I
- Meiosis II
- Phases of meiosis, values of c, n
- Karyotypes
- Sex chromosomes

3- Independent assortment
- Meiosis I
- Gametes formed during meiosis II
- Linkage: independent assortment of chromosomes not necessarily markers

4- Microscopy
- Mitosis versus meiosis
- "c" versus "n"
- haploid versus diploid

X-linked Inheritance
1- X-linked trait
- Dominant versus recessive
- X-linked versus autosomal
- Genotypes

2- X-linked trait (hemophilia)
- Genotypes
- Probability of affected children

3- X-linked trait (vitamin D rickets)
- Dominant versus recessive
- X-linked versus autosomal
- Genotypes
- Probabilities

4- X-linked trait (red-green colorblindness) and sickle-cell anemia (autosomal)
- Independent assortment
- Probabilities

5- Nondisjunction
- Colorblindness and Turner syndrome
- Colorblindness and Klinefelter syndrome
- Identify parent where nondisjunction occurred
- Nondisjunction in meiosis I or II
- White in flies
- Sex determination in flies

Fly Lab

1- Monhybrid cross (brown)
- Dominant versus recessive
- *Drosophila* nomenclature
- Genotypes
- Testcross
- Phenotypic ratios

2- Monohybrid cross (lobed)
- Dominant versus recessive
- *Drosophila* nomenclature
- Genotypes

3- Monohybrid cross (white)
- Reciprocal crosses to F_2
- Dominant versus recessive
- X-linked versus autosomal
- *Drosophila* nomenclature
- Genotypes
- Phenotypic ratios

4- Monohybrid cross (bar)
- Reciprocal crosses to F_2
- Dominant versus recessive
- X-linked versus autosomal
- *Drosophila* nomenclature
- Genotypes
- Phenotypic ratios

5- Dihybrid cross (ebony, purple)
- Dominant versus recessive
- X-linked versus autosomal
- Genotypes
- Phenotypic ratios
- Testcross
- Testcross ratios

6- Dihybrid cross (sable, dumpy)

- Reciprocal crosses
- Dominant versus recessive
- X-linked versus autosomal
- Genotypes
- Phenotypic ratios

Genotype/Phenotype

Genotype/Phenotype Problems

1- Heterozygote phenotype (snapdragons)
- One gene, two alleles, three phenotypes
- Genotypic ratios equal phenotypic ratios

2- Combination heterozygote phenotype (sickle-cell anemia)
- Analysis of proteins by gel electrophoresis
- Genotype determination from protein patterns
- Homozygosity versus heterozygosity
- Screens for heterozygosity
- Paternity testing

3- ABO blood types
- Two gene hypothesis versus one gene with multiple alleles
- Genotype determination

4- Multiple alleles (hypothetical birds)
- Multiple genes versus multiple alleles
- Allelic series
- Genotypes from phenotypic ratios

5- Two traits (hypothetical ducks)
- Breakdown of complex data
- Phenotypic ratios to predict number of genes and relationships between genes

6- Human example (Bombay phenotype)
- Genotypes
- Discrepancy in pedigrees
- Epistatic gene
- Southern analysis

7- Spontaneous mutation (curled ear in cats)
- Homozygous dominant never found
- 2:1 ratios
- Achondroplasia (human example)

Linkage

Linkage Problems

1- Dihybrid cross (corn)
- Linkage
- Recombination
- Cis arrangement
- Recombination frequency, map units
- Using F_2 progeny ratios to determine map units
- Using map units to determine F_2 progeny ratios

2- Dihybrid cross (forked and scalloped)
- Test cross versus sibling cross
- Recombinants and parentals
- Map distance

3- Dihybrid cross (forked and miniature)
- Cis versus trans
- Three two-factor crosses to build map (forked, scalloped, and miniature)

4- Dihybrid cross (spineless and incomplete)
- Test cross
- No recombination in males

5- Trihybrid cross (miniature, singed, tan)
- Gene order
- Genetic map

6- Dihybrid, trihibrid cross (mice)
- Genotypes
- Given map with three markers, predict number of F_2 progeny

Bacterial Genetics

Conjugation
Tutorials (on CD-ROM)
Conjugation
- Genetic exchange in bacteria
- F+ versus F- cells
- Animation of F+ x F-
- Integration of F+ (animation)
- Hfr strains
- Animation of Hfr x F-

Auxotrophy
- Biochemical pathways
- Gene mutation
- Replica plating video
- Antibiotic markers
- Sugar markers

Conjugation Problems
1- Selectable markers
- Amino acid markers
- Antibiotic markers
- Sugar markers
- Replica plates
- Genotypes

2- F+, F-, and Hfr cells
- Many, few, or no recombinants
- Predicting cell type from data

3- Interrupted mating
- Selection for exconjugants
- Time of entry
- Map in minutes from origin of transfer

4- Natural Gradient of Transfer
- Auxotrophic marker to select against Hfr parent
- Selection of two markers per plate
- Gene order using data

5- Multiple Hfrs
- Circular map

Transduction
> **Tutorial (on CD-ROM)**
>> **Transduction**
>>> - Phage life cycle (animation)
>>> - Transducing phage
>>> - Donor
>>> - Recipient
>>> - Co-transduction
>>> - Co-transduction frequency

> **Transduction Problems**
>> **1- Two-factor crosses**
>>> - Co-transduction with five auxotrophic markers
>>> - Co-transduction frequencies
>>> - Gene order
>>
>> **2- Three-factor crosses (markers in cis)**
>>> - Genotype frequencies from replica plates
>>> - Co-transduction frequencies
>>
>> **3- Simulation of three-factor cross (markers in trans)**
>>> - Choice of donor or recipient strain
>>> - Choice of selectable media
>>> - Replica plate data
>>> - Final map determination
>>
>> **4- Reciprocal crosses**
>>> - Two crossovers versus four crossovers
>>> - Gene order with respect to nearby marker

Biochemical Genetics

> **Tutorials (on CD-ROM)**
>> *Neurospora*
>>> - Asexual life cycle
>>> - Sexual life cycle
>>
>> **Gene-enzyme**
>>> - Beadle-Tatum experiment
>>> - Nutritional mutants
>>> - Complementation
>>> - Supplementation
>
> **Problems**
>> **1- Complementation groups**
>>> - Heterokaryons
>>
>> **2- Recombination**
>>> - Linkage analysis in *Neurospora*
>>> - Recombination versus complementation
>>
>> **3- Supplementation**
>>> - Supplementation
>>> - Biochemical pathway
>>
>> **4- *Saccharomyces cerevisiae***

- Yeast life cycle
- Complementation groups (gal system)
- Recombination measured in spores
- Genetic map

5- **Flies**
- Simulation of fly crosses
- Complementation groups in flies
- Recombination measured in flies

Population Genetics

Problems

1- Hardy-Weinberg
- Allele frequencies
- p and q by counting alleles
- Genotype frequencies
- Random sampling
- Hardy-Weinberg equation

2- Co-dominant alleles
- Determining p and q from population
- Determining if population is in HWE
- Chi-squared test

3- Marriages between populations
- p and q in the two populations
- Genotype frequencies in the progeny

4- Estimating heterozygote frequencies
- Common recessive trait (non-taster for PTC)
- Heterozygote frequency estimation
- Heterozygote frequency among dominant phenotype class

5- X-linked recessive
- Genotype frequencies: male versus female
- Proportion of marriages with particular outcomes

Molecular Markers

Tutorial (on CD-ROM)
Molecular markers
- RFLPs and Southern analysis
- VNTRs and haplotypes
- Single and multilocus probes
- PCR

Problems
1- Identification of linked RFLP
- Multiple restriction enzymes
- Multiple probes

2- Recessive alleles
Part a
- RFLP
- Southern
- Molecular marker linked to gene
Part b

- RFLP
- Southern
- Molecular marker linked to gene
- Probabilities
- Probabilities

3- Dominant or X-linked alleles

Part a (dominant)
- Linkage
- Probabilities

Part b (X-linked)
- Linkage
- Probabilities

4- Linked markers in yeast
- Linkage
- Map distance

5- Linked markers in humans
- Linkage
- Probabilities taking into account possible recombination

6- DNA fingerprinting
- STRP allele frequencies, genotype frequencies
- Probabilities

Medical Genetics

Tutorials (on CD-ROM)
Chromosomal disorders
- Trisomies
- Aberrations

Single gene disorders
- Inheritance patterns

Multifactorial disorders
Pedigree analysis
Karyotype analysis
Molecular techniques to detect mutant alleles
- ASO
- FISH
- PCR

Case Studies
- George Smith
- Han Chen and Yuh Nung Lee
- Rebecca Goldenstein
- Jan de Broek

Molecular Biology

Gene Expression
Tutorials (on CD-ROM)
Transcription
- Prokaryotes
- Eukaryotes

Splicing
- snRNPs and spliceosome formation
- transesterification 1 and 2

Translation
- Genetic code
- tRNA and ribosomes
- Intiation, prokaryotes, and eukaryotes
- Elongation
- Termination

Problems

1- Promoters
- Alternative sigma factors
- Consensus sequence
- Start site
- 5' end
- Chemical reaction
- Template strand
- Promoter mutants

2- Alternative splicing
- Calcitonin versus CGRT
- Southern analysis
- Splice sites and alternative splicing

3- Hemoglobin mutants
- SDS versus native gel (Western analysis)
- Northern versus Southern
- Various translation mutants

Gene Cloning

Tutorials (on CD-ROM)

Restriction enzymes
- Palindromes
- Sticky versus blunt ends
- Frequency
- Ligation
- Methylation

DNA sequencing
- Nucleotide addition
- Inhibition by dideoxynucleotides
- Animation of DNA sequencing

Southern blot
- Restriction digest of total DNA
- Transfer of fragments to nitrocellulose (video)
- Radioactive probe
- Autoradiogram

PCR
- PCR animation

Problems

1- Restriction mapping
- Single and double digests
- Gel electrophoresis

- Size of DNA
- Equations
- Restriction map

2- Cloning hemoglobin gene from patient
- Splicing defect indicated
- Wild-type gene available
- Choice of insert, vector, method of identification

3- Cloning DNA polymerase gene
- Given pure protein
- Choice of insert, vector, method of identification

4- Cloning polydactyly gene
- Given linked RFLP
- Choice of insert, vector, method of identification

Mendelian Analysis
....

Goals for Mendelian Analysis:

1. Describe the mode of inheritance of a phenotypic difference between two strains:
   - Distinguish dominance and recessiveness in traits
   - Determine whether the phenotype difference is due to a single gene with two alleles
   - Write genotypes and predict phenotypes
   - Predict the number of possible gamete types, their kinds and ratios
2. Distinguish self-crosses from test crosses.
3. Apply these concepts to simple human pedigrees for select traits:
   - Calculate probabilities using product and sum rules
   - Use the binomial expansion to calculate the combinations of possible outcomes

Mendelian Problems
Problem 1

Let's begin with a simple cross between two pure breeding peas, a purple flowered pea and a white flowered pea. How many traits or characters are different between the purple and white parents?

In this cross the pollen from the purple parent is used to fertilize the ovules of the white parent. How many genetically distinct gametes are produced from the purple parent?

How many genetically distinct gametes does the white flowered parent produce?

All the progeny of the parental cross are purple flowered pea plants. However, when these peas are self-crossed, both purple and white peas appear in the next or F_2 generation. Which phenotype is the dominant phenotype?

Using A and a to designate the different forms of the inherited trait (alleles), assign genotypes to the three generations of the cross described above. How many genetically distinct gametes are produced from the F_1? In the simple Punnett square shown below, fill in the genotype of the progeny using A/a.

gamete types

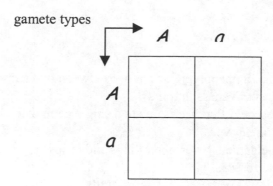

Indicate the phenotype expected for each genotype in the Punnett square above. What is the ratio of purple flowered progeny to white flowered progeny in the F_2?

What genotypic ratios would you predict?

Problem 2

The grass tree is the common name for the Australian genus *Xanthorrhoea* in the lily family, *Liliaceae*. Grass trees grow to a height of about 4.5m (15 feet) and bear long, narrow leaves in a tuft at the top of the trunk. White flowers or yellow flowers can be produced in a dense spike above the leaves in a given tree. A tree producing yellow flowers was self-crossed. The seeds were collected and planted to determine flower color of each progeny tree. Twenty-eight trees were found to produce yellow flowers and eight trees were found to produce white flowers.

What is the expected genetic (phenotypic) ratio of yellow flowering grass trees to white flowering grass trees among the progeny?

Which phenotype is dominant, white or yellow flowering trees?

Using the symbols of *Y/y*, write the genotypes of the original yellow flowering tree and the progeny.

Yellow × Yellow

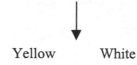

Yellow   White

Predict the phenotypic ratios for the crosses shown below:

Cross	Phenotypic Ratios	
	Yellow	White
White tree × white tree		
Original yellow tree × white tree		
Pure breeding yellow tree × white tree		
Original yellow tree self-crossed		

Problem 3

Two pure breeding strains of peas, one giving wrinkled, yellow seeds and the other round, green seeds, were crossed and all the resulting peas were round and yellow. These round seeds were then planted and the flowers self-fertilized. The peas produced from this selfing contained four phenotypic classes.

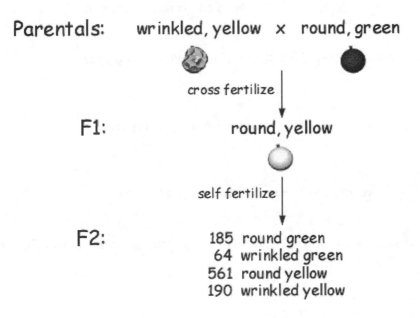

Parentals: wrinkled, yellow × round, green

cross fertilize

F1:   round, yellow

self fertilize

F2:   185 round green
            64 wrinkled green
           561 round yellow
           190 wrinkled yellow

How many traits or characters are different between the pure breeding parents?

Using A and a to designate pea shape and B and b to designate pea color, assign genotypes to the three generations in the diagram of the cross above.

How many genetically distinct gametes are produced from the F_1?
Fill out the Punnett square shown below.

gamete types

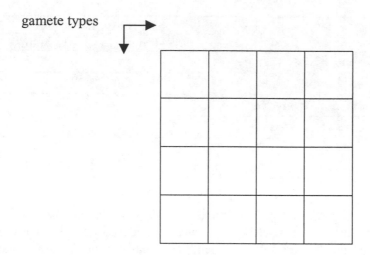

How many genotypes above correspond to the round and yellow phenotype?

How many genotypes above correspond to the wrinkled yellow phenotype?

How many genotypes above correspond to the round green phenotype?

How many genotypes above give rise to a green wrinkled pea?

Using the Punnett square above, what is the ratio of round peas to wrinkled peas (ignore the color)?

What is the ratio for yellow versus green?

Determine the ratio of round yellow peas to round green peas to wrinkled yellow peas to wrinkled green peas expected from this dihybrid cross.

Problem 4

You have been sent to Planet TATACCA to conduct basic genetic experiments on a new species: **X**. The two traits that you are observing are the shape and attachment of the ear. Pointy ears are dominant to round, and attached ear lobes are dominant to detached.

Assuming that genes responsible for these phenotypes assort independently, write out the genotype of the parents in each of the crosses. Assume homozygosity unless there is evidence otherwise. **P** and **R** stand for the pointy and rounded phenotypes, respectively, and **A** and **D** stand for the attached and detached phenotypes. Use *P* and *p* for the pointy and round ear alleles and *A* and *a* for the attached and detached.

Parental Genotype	Parental Phenotype	Number of Progeny			
		PA	*PD*	*RA*	*RD*
	PA × PA	447	152	147	50
	PA × PD	107	98	0	0
	PA × RA	95	0	91	0
	RA × RA	0	0	153	48
	PD × PD	0	149	0	51
	PA × PA	140	46	0	0
	PA × PD	151	147	48	50

Problem 5

Geneticists working with the common fruit fly, *Drosophila melanogaster*, are studying the mode of inheritance of vestigial wings. Vestigial wings are a mutant phenotype in which wing development is curtailed. Wild-type or normal flies display normal wing development. Geneticists working with flies label genes and the alleles, and therefore the flies, after the mutant phenotype. Pure breeding vestigial flies have the genotype of *vgvg* (two copies of the mutant allele) and the phenotype is denoted as *vg* for vestigial. Pure breeding normal flies are denoted vg^+vg^+ for their genotype and their phenotype is described as vg^+.

After sorting flies for several hours, a freshman volunteering in the laboratory accidentally sneezes, mixing up all three piles of flies. Originally, one pile of flies displayed vestigial wings, a second pile of flies had normal wings but was heterozygous, and the third pile had normal wings and was homozygous. The principal investigator of the laboratory calls on you, as an expert geneticist, to resort the flies. By performing several crosses, you are able to determine from which pile each fly originated.

Which phenotype is recessive, vestigial wings or wild-type wings? (See the table below).

Using the symbols of *vg+* and *vg*, write the genotypes of the flies used in each cross in the table below.

Genotype	Phenotype	Number of Progeny	
		vg+	*vg*
	normal × normal	628	209
	vestigial × normal	333	340
	normal × vestigial	454	0
	vestigial × vestigial	0	121
	normal × normal	92	0

Problem 6

One day as you were traveling along a semi-deserted highway, your car has an electrical surge and suddenly stops. You can't get your car to run again so you decide to get out and walk to the nearest phone to call for a tow truck. While you are walking you see a dairy farm in the distance. As you get closer you notice that these are not ordinary cattle. Then, you see the signs indicating a government research facility. Because you are extremely interested in the genetics of these odd cows, you take a few with you as you run off. Amazingly, you and your new bovines have made it to your family's farm safely. You decide to breed the cattle to obtain a better understanding of the genetic basis for their coat color. The results from your crosses are found on the table below.

Parental Genotypes	Parental Phenotypes	Green Progeny	Purple Progeny
	purple × purple	0	98
	green × green	94	0
	purple × green	51	49
	purple × green	97	0
	green × green	71	27

Which color results from possessing a dominant allele?

Using *G* and *g* as allele designations in the form of *Gg* × *Gg*, assign the most probable genotypes for the parents in each cross above. Assume homozygosity unless there is evidence otherwise.

Besides having either a purple or a green coat, your cattle have two other odd traits that you have observed to be inherited in a Mendelian manner. Your new bovines have either yellow eyes (*Y*) or blue eyes (*B*) and produce either regular white milk (*W*) or chocolate milk (*C*). Consider the table below. Which is the recessive eye color?

What type of milk is produced when a cow possesses a dominant allele?

Parental Phenotypes	Blue Chocolate	Yellow Chocolate	Blue White	Yellow White
BC × *BC*	10	4	3	1
BW × *YC*	7	8	0	0
BC × *BW*	4	0	3	0
BW × *BW*	0	0	15	5
YC × *YC*	0	22	0	7
BC × *BC*	12	4	0	0
BC × *YC*	10	11	3	4

Using *C* to indicate the allele for chocolate milk production, *c* for white milk production, *B* for blue eyes, and *b* to indicate the allele for yellow eyes, write the parental genotypes for each of the crosses shown above. Assume homozygosity unless there is evidence otherwise. What proportion of the offspring of two parental cattle each of the genotype *Gg Bb Cc* (green with blue eyes and chocolate milk producing) will be *Gg bb cc*?

Consider the following cross: *Gg Bb Cc* × *gg Bb cc*

What fraction of the progeny do you expect to phenotypically resemble the first parent?

If 100 progeny resulted from crosses such as this, how many would you expect to exhibit new genotypes (i.e., do not genotypically resemble either parent)?

Pedigree and Probabilities
Problem 1

Part a. For the pedigree shown below, state whether the condition depicted by darkened symbols is dominant or recessive. Assume the trait is rare. Assign genotypes for all individuals using *A/a* designations.

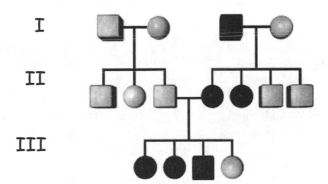

Part b. Assume the inherited trait depicted in the pedigree below is rare and state whether the condition is dominant or recessive. Assign genotypes for all individuals using *A/a* designations.

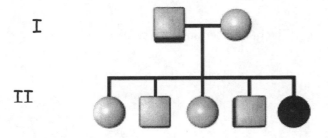

Part c. Is the pedigree shown below consistent with a dominant or recessive trait?

In analyzing this pedigree, would you conclude that this trait is caused by a rare or common allele?

Assign allele designations using *A/a*.

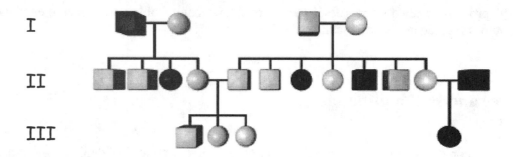

Problem 2

You have a single six-sided die. What is the chance of rolling a six?

What is the chance you will roll a five or a six? (HINT: Should you use the product rule or the sum rule to determine the probability of rolling one or the other?)

You are now given two additional dice. If you roll all three dice simultaneously, what is the chance of obtaining a five on all three dice?

Rolling your three dice again, what is the chance of obtaining no fives at all?

What is the chance of obtaining two fours and one three on any of the dice in a single roll?

What is the chance of obtaining the same number on all three dice?

What is the chance of rolling a different number on all three dice?

Problem 3

Preparing for a long night studying genetics, you open a big bag of candy-coated peanuts. Two friends are supposed to join you so, so you divide the bag into thirds. As you wait, you get hungry. But since you want to keep the bowls equal, you eat one from each bowl.

Bowl 1	Bowl 2	Bowl 3
30 blue candies	40 blue candies	30 blue candies
30 yellow candies	50 yellow candies	20 yellow candies
40 brown candies	10 green candies	50 red candies

What is the chance that you will select three blue candies?

What is the probability of selecting a brown, a green, and a red candy?

If you pick one candy from each bowl, what is the chance you will pick two blue candies and a yellow candy?

If you pick one candy from each bowl, what is the probability of obtaining no yellow candies?

If you pick one candy from each bowl, what is the chance of picking at least one yellow?

The probability of having two peanuts in a single candy is 2%. If you eat 300 candies, how many double peanuts do you expect to find?

Problem 4

The ability to taste the chemical phenylthiocarbamide (PTC) is an autosomal dominant phenotype. Lori, a taster woman, marries Russell, a taster man. Lori's father and her first child, Nicholas, are both nontasters.

What is the probability their second child will be a nontaster girl?

What is the probability their second child will be a taster boy?

What is the probability that their next two children will be nontaster girls?

Problem 5

A rare recessive allele inherited in a Mendelian manner causes phenylketonuria (PKU), which can lead to mental retardation if untreated. Fortunately, with a diet low in phenylalanine and tyrosine supplementation, normal development and lifespan are possible. Mike, a phenotypically normal man whose father had PKU marries Carol, a phenotypically normal woman whose brother had the disorder. The couple wants to have a large family and come to you for advice on the probability that their children will have PKU.

What is the probability that the couple's first child will have PKU?

Their first child, Greg, has PKU and the couple wants to have five additional children. What is the probability that out of their next five children only two will have PKU?

What is the possibility that they will have no more than one affected child in the five additional children they are planning?

Problem 6

John and Maggie are expecting a child. John's great grandmother (mother's lineage) and Maggie's brother have a rare autosomal recessive condition. What is the chance that their child will be affected?

John and Maggie have just discovered they are going to have twins. What is the chance that both twins will be affected if they are identical twins?

If the twins are dizygotic twins (non-identical or two-egg twins), what is the chance they will both be affected?

If they are dizygotic twins, what is the chance that at least one of them will be affected?

Chromosomal Theory of Inheritance

Goals for Chromosomal Inheritance:

1. Understand the classical evidence that genes are located on chromosomes:
   - Similarity of behavior of chromosomes in divisions with the behavior of genes (alleles) in inheritance
   - Identification of genes with sex-linked inheritance and chromosomes with sex-linked inheritance
2. Distinguish mitosis from meiosis figures, as well as the species diploid chromosome number, by inspection of simple diagrams with chromosome size, shape, and copy number as the cues.
3. Connect genetic inheritance with chromosome behavior during divisions:
   - Assign alleles to chromosomes: sister chromatids have identical alleles, homologous chromosomes (in a heterozygote) have different alleles
   - Homologous chromosomes pair and disjoin from each other in the first meiotic division, illustrating segregation of the alternate alleles
   - Sister chromatids separate at mitosis, illustrating the constancy of the genotype in both daughter cells
   - Genes that are on different chromosome pairs, i.e., non-homologous pairs, always assort independently
4. Be able to identify sex-linked inheritance in a pedigree.

Mitosis and Meiosis
Problem 1

Many cells undergo a continuous alternation between division and nondivision. The interval between each mitotic division is called interphase. Which stages (G1, S, G2 or M) make up interphase?

It was once thought that the biochemical activity during interphase was devoted solely to cell growth and specific functions the cell normally performs. However, it is now known that another biochemical step critical to the next mitosis occurs during interphase: the replication of the DNA of each chromosome. During which stage of the cell cycle is DNA replicated?

The concentration of DNA contained in a single set of chromosomes is frequently referred to as "c." Diploid somatic cells have two sets of chromosomes and alternate between 2c and 4c. A cell has 2c before DNA replication. Then, it has twice that amount, 4c, until it undergoes mitotic division. After mitosis, the two resulting cells will both have 2c.

The job of the mitotic division phase is to segregate the replicated chromosomes into two cells, each with the same chromosome and genetic complement as the parent cell. The four phases of mitosis are: prophase, metaphase, anaphase, and telophase.

DNA replication (in S phase) produces a second double helix. During prophase, the sister double helices condense, forming the X shaped chromosomes typically seen under the microscope.

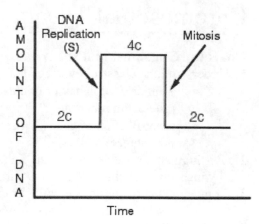

Draw a cell containing one pair of homologous chromosomes after it undergoes DNA replication (i.e., S phase) as seen in metaphase.

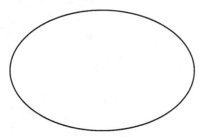

Suppose one homologue carries the *a* allele and the other homologue carries the *A* allele. Label each chromatid in your drawing.

Put each of the pictures below (containing two pairs of homologous chromosomes) in correct order and label with the appropriate phase. Note: Homologous chromosomes do not have to be next to each other during mitosis.

    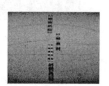

_____   _____   _____   _____   _____

Draw two daughter cells generated by a mitotic division with the appropriate homologues (labeled with *A* and *a*).

Problem 2

The results Mendel observed from his early crosses established the idea that alleles segregate. Unfortunately, Mendel couldn't figure out the biological mechanism behind this principle.

About 100 years ago, scientists realized that gametes (sex cells) were the result of a specialized cell division. This division process began with a cell that had two sets of chromosomes (a diploid cell) and ended with four cells with only one set of chromosomes (a haploid gamete). Today we know that the segregation of homologous chromosomes within this division process, meiosis, provides the mechanism for the segregation of alleles.

During sexual reproduction, two gametes (one from the mother, one from the father) combine in fertilization to form a diploid cell. A diploid fly has eight chromosomes. Without meiosis, how many chromosomes would the first generation progeny contain?

The purpose of meiosis is to convert one diploid cell into four haploid cells, each containing a single complete set of chromosomes. Meiosis can be divided into two cycles: Division I and Division II (or meiosis I and meiosis II). Two cells result from meiosis I. For a cell with a single pair of homologous chromosomes (labeled with *A/a*), draw the chromosomes found in the two cells formed by the first meiotic division. Then draw the four products of meiosis, with the appropriately labeled chromosome.

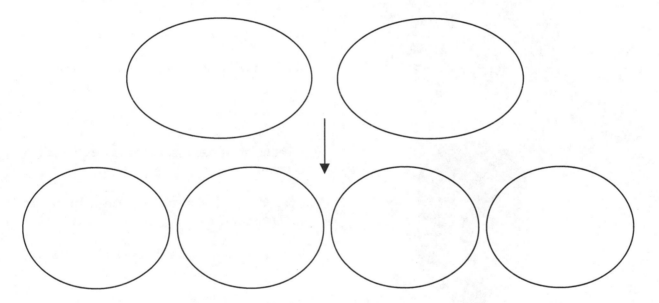

How many types of gametes with different alleles are formed in this example?

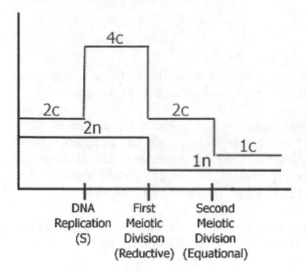

Regardless of whether there are one or two chromatids per chromosome, n indicates the number of chromosomes in a cell. Recall that c refers to the concentration of DNA in a cell.

a

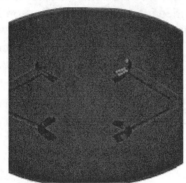

What phase of meiosis best describes the cell in Figure a?

How many pairs of homologous chromosomes are there in this cell?

How many chromatids are present?

What is the value of c at this stage of meiosis?

b

What phase of meiosis best describes the cell in Figure b?

Does the cell shown here have 1n or 2n chromosomes?

Is the amount of DNA best described as c, 2c, or 4c?

c

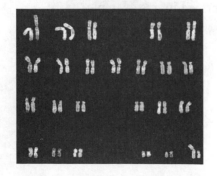

A karyotype (Figure c) is the ordered visualization of a complete set of chromosomes from metaphase. Karyotypes can be used to determine the number of chromosomes a species has, as well as any abnormalities an individual chromosome might have. This is the karotype of a male gorilla. How many chromosomes does a gorilla have? How many chromosomes are found in a gorilla gamete?

Autosomes are the same in both males and females. Sex chromosomes (bottom right corner) differ according to sex (XX in females, XY in males). X and Y chromosomes behave as homologues in males (the X and Y chromosomes pair and segregate at meiosis I). How many chromosomally different gamete types does a male produce?

Problem 3

Through the analysis of dihybrid crosses, Mendel was able to deduce that the genes he was studying were assorting independently, giving rise to gametic ratios of 1:1:1:1. However, in 1865, the physical nature of this independent assortment was not understood. After the discovery of meiosis in the 1880s, scientists recognized that genes located on different chromosomes should assort independently. A physical basis for Mendel's hypothesis was now possible!

The genetic material must be duplicated (or replicated) prior to its assortment. Draw two pairs of homologous chromosomes (labeled with *A/a* and *B/b*) in the cell below as they would appear after DNA replication (in prophase I).

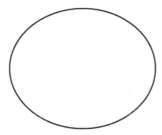

Two cells result from the first division of meiosis. Draw the chromosomes as they would appear in the cells after meiosis I. Remember that there are two different, equally probable ways for the chromosomes to assort (just draw one).

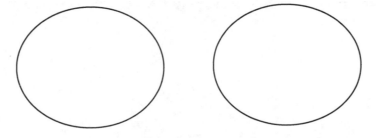

Write the possible gametes produced by a female with the genotype *AaBb*.

What are the possible gametes produced by a male with the genotype *AaBb?*

Notice that the gametes produced are identical to the meiotic products for an *AaBb* × *AaBb* cross. Recall that the Punnett square accurately predicts the genotypic frequencies when the genes involved assort independently of one another. In further testing Mendel's law of independent assortment, you cross an *Aa Bb Dd* female mouse with an *aa bb dd* male. How many possible gamete types can the female produce?

Suppose instead of seeing the number of classes of mice you expected, you find only four. You find 30 *abd*, 33 *aBd*, 29 *AbD*, and 28 *ABD* mice. In order to explain this enigma, we will look at one possible arrangement of genes on the mother's chromosomes.

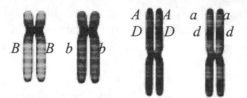

Two cells result from the first division of meiosis. Draw the chromosomes shown above to create the cells after meiosis I. Note that even though the two chromosomes are assorting randomly, the *A* and *D* alleles travel together since they are on the same chromosome.

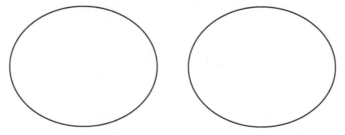

Completion of meiosis will produce four gamete types. What are the four possible genotypes (consider both possible assortments)?

This is an example of tightly linked genes on a chromosome where no crossing over can occur (see topic Linkage).

Problem 4

Problem 4 is a series of cartoons of cells at different stages of meiosis and mitosis and should be done on the CD-ROM.

X-Linked Inheritance
Problem 1

Is the pedigree below consistent with a dominant or a recessive trait? Assume the trait is rare.

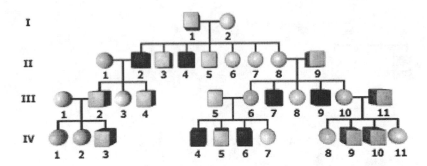

Is this pedigree consistent with X-linked or autosomal inheritance? Assume the trait is rare.

In the pedigree above, is there any evidence of father to son transmission of the trait?

What is the genotype of individual II-2? Write *A* or *a* for the X-linked alleles and Y for the Y chromosome.

What is the genotype of individual III-6? Write *A* or *a* to designate the X-linked dominant or recessive alleles.

Suppose in the pedigree above, individual II-8 was a male instead of a female and individual II-9 a female instead of a male. All other individuals are unchanged. Would this change the mode of inheritance deduced from this pedigree?

Is this pedigree still consistent with a recessive trait?

Can you tell whether the trait, in this hypothetical pedigree, is common or rare?

Now that the mode of inheritance is different, what is the genotype of individual II-2? Use *A* or *a* for the dominant and recessive alleles, respectively.

Problem 2

Hemophilia is a rare, recessive X-linked disease that usually affects only males. Consider the pedigree below of a family of normal parents with a son who has hemophilia.

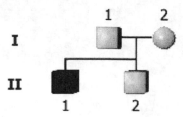

What is the father's genotype? Write H for X^H, h for X^h, or Y.

What is the mother's genotype? Write H for X^H and h for X^h.

They are going to have another child, whom they know is male (individual II-3). What is the chance that he will be affected?

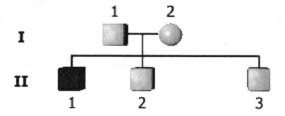

What is the chance that a daughter (individual II-5) will be a carrier?

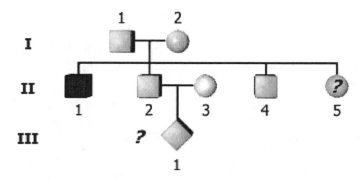

What is the probability that the Child III-1 will be affected?

What is the probability that the Child III-1 will be a carrier girl?

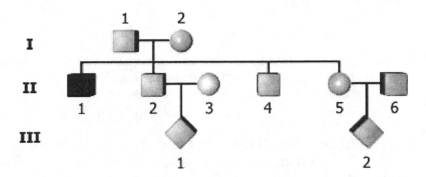

What is the probability that the Child III-2 will be an affected boy?

Problem 3

The pedigree below belongs to a family with a rare trait known as vitamin D resistant rickets. Is this pedigree consistent with a dominant or recessive mode of inheritance?

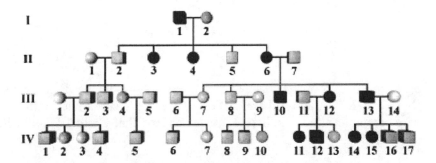

Is this pedigree consistent with an autosomal or an X-linked trait?

Assuming the trait is X-linked, what is II-6's genotype? Use *A* or *a* to designate the dominant or recessive X-linked alleles.

What is the genotype of individual II-5? Use *A* or *a* to designate the dominant or recessive X-linked alleles.

Suppose that III-14 has just found out she is going to have another girl. What is the probability that the child will be affected?

Individuals III-11 and III-12 are going to have another child but they don't know its sex. What is the probability they will have an affected son?

Problem 4

Red green colorblindness is a common X-linked recessive trait in humans that affects both males and females. Affected individuals are unable to distinguish red from green. Its prevalence accounts for the fact that you see some affected females as well as affected males.

Sickle cell anemia is an autosomal recessive disorder with a high incidence in the African American population (1/400). It is a structural hemoglobin abnormality that causes sickling of red blood cells with resulting complications.

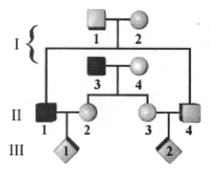

Consider this pedigree of two brothers marrying two sisters. All the individuals in generation II, namely II-1, II-2, II-3, and II-4 are carriers of the sickle cell allele. Individuals with filled blue squares are colorblind. Is there a greater chance that Child III-1 will have sickle cell anemia than Child III-2?

What is the chance that Child III-1 will be colorblind and have sickle cell anemia?

What is the probability that Child III-2 will be colorblind and affected with sickle cell anemia?

If the child (III-2) from the second marriage is colorblind, what is the child's gender?

Problem 5

Nondisjunction is a rare event that occurs in meiosis when paired chromosomes or sister chromatids do not disjoin properly. When this occurs in meiosis I, both homologues end up in one daughter cell. When nondisjunction occurs in meiosis II, one of the four meiotic cells ends up with both sister chromatids from one chromosome, and the other meiotic cell involved doesn't have any copies of that chromosome.

In humans, nondisjunction is generally lethal except when it involves chromosome 21 (leading to Down syndrome) or the sex chromosomes. In the following examples, nondisjunction of the sex chromosomes will be examined in conjunction with the X-linked gene that causes red green colorblindness. Affected individuals are unable to distinguish red color from green. In severely affected individuals, everything appears gray.

A man who is colorblind has a daughter with Turner syndrome who is also colorblind. You want to explain the origin of this daughter through nondisjunction in one parent.

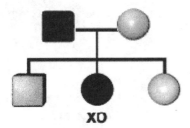

Assuming the mother is not a carrier of the colorblindness allele, use the genetic marker of colorblindness in this family to determine which parent had the nondisjunction event.

Can you tell at what stage of meiosis (meiosis I or meiosis II) nondisjunction occurred?

A woman who is colorblind has a son with Klinefelter (XXY) syndrome who is not colorblind. You want to explain the origin of this son through nondisjunction in one parent. Start by assigning a genotype to each individual.

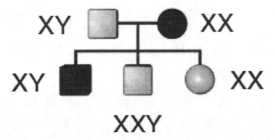

Can you use the genetic marker of colorblindness in this family to determine which parent had the nondisjunction event? Indicate that parent in the pedigree above.

Can you determine whether the nondisjunction event occurred in meiosis I or meiosis II of the parent's gametic line?

In the *Drosophila* fruit fly, the gene for white eyes is located on the X-chromosome. The white-eyed allele is recessive to the wild type. In crosses between white-eyed females and wild-type males almost all (regular) daughters have red eyes, except for about 1/2000 (0.05%) exceptional white-eyed daughters. Almost all (regular) sons have white eyes except for about 1/2000 exceptional red-eyed sons that are sterile.

Upon inspection of the flies' karyotypes, the white-eyed exceptional daughters were found to carry XXY. (Regular females are XX and males are XY.) Does this finding indicate that the Y chromosome determines sex in fruit flies?

Indicate the correct alleles of the exceptional females, one allele at a time. Follow the *Drosophila* nomenclature rules (see Fly Lab below) and choose between *w* and *W* for the white-eyed allele, and between *w+* and *W+* for the wild-type allele.

When the exceptional red-eyed sons are inspected, they are found to be XO, having a single X chromosome. Does this fact agree with the conclusion that the presence of Y does not determine sex in the *Drosophila* fruit flies?

Do these unusual findings of white-eyed exceptional daughters and red-eyed exceptional sons represent nondisjunction in the *Drosophila* father or *Drosophila* mother?

Can we tell whether the nondisjunction occurred in the first meiotic division or in the second meiotic division?

Fly Lab

Welcome to the fly lab! The following problems allow you to play the role of a geneticist working with the common fruit fly, *Drosophila melanogaster*. Since *Drosophila* nomenclature may be confusing at first, here is a brief review:

Genes are named after the mutant phenotype in *Drosophila*. Therefore, if the mutant gene results in the lack of eye formation, the gene could be named "eyeless." (Note that this differs from the Mendelian designations used for peas, for example, where the gene is typically named after the dominant trait.)

When allele symbols are assigned, lower case is used if the mutant allele is recessive (for example, *ro* for rough eyes). Capitalizing the first letter indicates that the mutant allele is inherited as a dominant trait (for example, *N* for notched wings).

Wild-type alleles are designated with a superscript + (eg., ro^+ or N^+). In the following problems simply indicate the + after the allele symbol (i.e., *ro+* or *N+*).

The problems below are set up as simulations on the CD-ROM. Here, the results are provided.

Problem 1

In *Drosophila*, the brown mutant is characterized by a brown eye color compared with the brick red, wild-type color. A wild-type fly is crossed with a brown-eyed fly to produce the F_1 generation and then the F_1 flies are crossed to produce the F_2 generation.

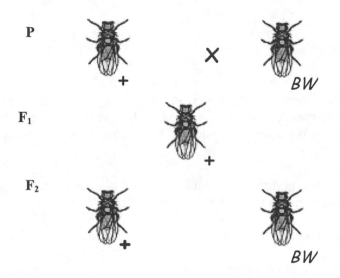

Is the brown mutation dominant or recessive?

Using the letters *BW* and proper Drosophila nomenclature (*bw* for recessive and *Bw* for dominant), indicate the proper mutant allele designation below.

What is the genotype of each of the flies above?

Problem 2

In *Drosophila*, the lobed mutation is characterized by smaller eyes compared to the wild type. A wild-type fly is crossed with a lobed fly to produce the F_1 generation (all lobed). Then the F_1 flies are crossed with each other to produce the F_2 generation (768 lobed and 259 wild type).

Is the lobed mutation dominant or recessive?

How would you indicate the mutant allele using the correct *Drosophila* nomenclature?

What is the genotype of each of the parental, F_1, and F_2 flies?

What fraction of the mutant F_2 flies is homozygous?

Problem 3

In *Drosophila*, the white mutant is characterized by white eyes compared to the brick red, wild-type color. A wild-type female fly is crossed with a white-eyed male to produce 974 wild-type F_1 flies. At this point what can you conclude about the mutation (is it dominant or recessive)?

The reciprocal cross is then performed: A wild-type male is mated with a white-eyed female to produce 436 white-eyed males and 421 wild-type females.

What can you now conclude about the mode of inheritance (autosomal or X-linked)?

Now go back to each of the crosses above and write the genotype of all the flies.

Problem 4

In *Drosophila*, the bar mutant is characterized by eyes that are restricted to a narrow, vertical bar. When a bar female is mated to a wild-type male, all the F_1 flies are bar. However, when a bar male is mated to a wild-type female, 857 bar females and 905 wild-type males are observed.

What is the mode of inheritance of the bar mutant?

What is the genotype of each of the flies in the two crosses above?

Problem 5

In *Drosophila*, the ebony mutant is characterized by an ebony body color and purple is characterized by purple eyes. Mating an ebony, purple female with a wild-type male yields all wild-type progeny. The reciprocal cross gives the same results.

What is the mode of inheritance for the ebony mutation?

What is the mode of inheritance for the purple mutation?

What is the genotype of the F_1 flies?

Mating the F_1 flies together yields 226 wild type, 74 ebony, 78 purple, and 25 ebony, purple flies. What is the ratio of progeny for each of the phenotypic classes?

Which F_2 fly should you use for a testcross of the F_1 flies?

How many different phenotypic classes, and in what ratios, do you expect from this cross?

Problem 6

In *Drosophila*, the sable mutant is characterized by a sable body color and dumpy is characterized by shorter, oblique wings. In a cross between a sable, dumpy female and a wild-type male, all the female progeny are wild type and the male progeny are sable. When the F_1 siblings are mated, the F_2 consists of 338 wild type, 336 sable, 114 dumpy, and 110 sable, dumpy (both male and female).

What is the mode of inheritance of the sable mutant?

What is the mode of inheritance of the dumpy mutation?

In order to understand the unusual ratios and the lack of apparent linkage to sex of the sable phenotype in the F_2, assign genotypes to the parental and F_1 generations. Write the genotypes above, using s or $s+$ or Y to indicate sable alleles and dp or $dp+$ to indicate dumpy alleles. Predict the proportion of each phenotype you would expect in the F_2.

Genotype/Phenotype

Goals for Genotype/Phenotype:

1. Understand the following phenomena that lead to variations on Mendelian phenotypic ratios:
   - Incomplete dominance
   - Codominance
   - Epistasis
   - Homozygous lethality
2. Recognize combinations of 9:3:3:1 phenotypic ratios, where the phenotype comes from two genes and involves epistasis.
3. Distinguish from phenotypic ratios whether multiple genes or a single gene with multiple alleles are involved in determining the phenotypes.

Problem 1

Gardeners at the Japanese Botanical Garden discovered that after planting only red and ivory snapdragons some plants with pink flowers appeared among the progeny. The gardeners decided to experiment with the snapdragon flowers and carried out a number of additional crosses. In this diagram, the parents are shown along the margins with progeny types inside the boxes.

Parents	Red	Ivory	Pink
Red	red	pink	red, pink
Ivory	pink	Ivory	ivory, pink
Pink	red, pink	pink, ivory	red, ivory, pink

Which of the parents are homozygous?

Which of the parents are heterozygous?

By crossing the pink plants we should be able to distinguish whether their color is caused by one or more gene differences between the red and ivory parents. Take a moment to predict what offspring you expect for one gene versus two genes.

From the original results, the pink × pink cross yielded three phenotypes among the progeny. The numbers of each of the phenotypes are listed below.

F_2: 261 red
489 pink
243 ivory

What is the ratio suggested by these results?

How many <u>genes</u> differ between the red and ivory parents in determining flower color?

Assume the gene differing between the red and ivory parents has two alleles, *P1*, associated with the presence of red pigment, and *P2*, associated with the absence of red pigment. What are the genotypes of the different progeny types in the F_2 above?

Which of the following concepts best explains the observed results of the F_1 and F_2 progeny: incomplete dominance, codominance, *P1* dominant, *P2* dominant, multiple alleles, or multiple genes?

Problem 2

The tools we use to infer genotypes are phenotypes and the results of crosses. A powerful advance is the ability to look at phenotypes that are closer to gene activity. Here we describe an example using gel electrophoresis of proteins.

Proteins can be separated based on size and/or charge using electrophoresis through a gel matrix. Migration distance through the gel serves as a phenotypic marker that could help determine an individual's genotype. Such is the case with the normal and the sickle cell hemoglobin molecules. With sickle cell, heterozygous individuals produce both normal and sickle cell hemoglobin molecules. When the normal and the abnormal hemoglobin molecules are run through a gel, they migrate at different rates and can be visually separated.

Part a. The family below has a history of sickle cell anemia. The hemoglobin electrophoresis pattern for each child is shown in the lane below that child.

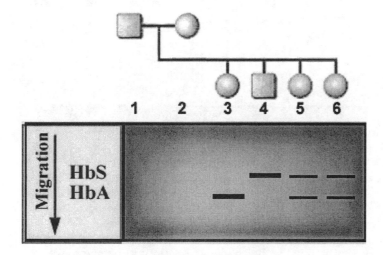

Which individuals in this family are homozygous for the sickle cell allele?

Which individuals from this family are heterozygous?

Imagine that this family is one out of a large number of families with heterozygous parents tested for their hemoglobin structure. Predict the ratios of children expected to be homozygous for *HbA*, homozygous for *HbS*, and heterozygous.

Which of the following concepts can best explain the expected phenotypic ratio: *HbA* dominant, *HbS* dominant, incomplete dominance, or codominance?

Part b. Gel electrophoresis can be used to screen individuals for the *HbS* allele.

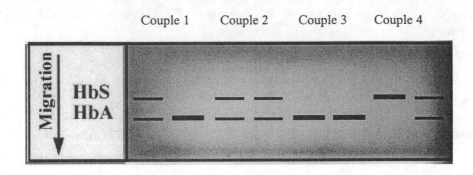

Which of the couples above are at risk of having an affected child?

What is the probability that Couple 2 will have a child with sickle cell anemia?

Part c. Use the Southern blot below to determine which of the three males (4, 5, or 6) could be the children's father.

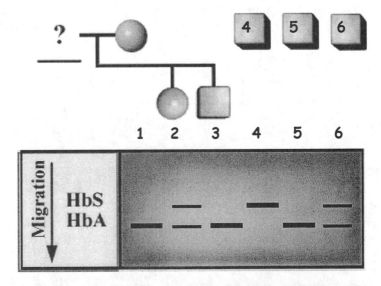

If this couple has another child, what is the chance the child will be anemic?

Problem 3

Karl Landsteiner was an Austrian-American physician who discovered that human blood differed in the capacity of serum to agglutinate red blood cells. By 1902, he and his group divided human blood into the groups A, B, AB, and O. Landsteiner concluded that two genes, A and B, control the ABO blood system he discovered. He proposed that each gene had two alleles, the presence and the absence of that allele.

Genotype	Phenotype
$A-bb$	A
$aa\ B-$	B
$A-B-$	AB
$aa\ bb$	O

Since you have been volunteering in Labor and Delivery for quite some time, you decide to compare the blood types you would expect using Landsteiner's two-gene hypothesis to the blood types you have observed. What genotype do you expect the children of two type O parents will have?

What is the expected phenotype of their children?

From your observations, 500 O × O parents have 763 type O children, whereas no type A, type B, or type AB children are observed. Is this consistent with the two-gene hypothesis?

One day you realize that in all 503 O × AB couples you have never seen any AB or O children. You have observed 600 type A and 650 type B children. Is this what you would predict based upon Landsteiner's hypothesis?

*Additional related questions are on the CD-ROM.

Problem 4

Suppose you are studying a novel bird species that displays a variation in feathers (blue, green, teal, and purple). Starting with pure breeding males and females from each phenotypic class, you perform the crosses diagrammed below.

Parents	F_1	F_2
teal × green	teal	¾ teal, ¼ green
green × blue	green	¾ green, ¼ blue
teal × purple	teal	¾ teal, ¼ purple
green × purple	green	¾ green, ¼ purple
blue × purple	purple	¾ purple, ¼ blue

Based on the data shown above, is feather color in this novel species segregating as if it were associated with multiple genes or multiple alleles?

Using the data shown above, which allele is dominant, blue or green?

Blue or purple?   Green or purple?   Green or teal?

What is the order of alleles corresponding to increasing dominance?

Cross	Parental Phenotypes	Phenotypes of Progeny			
		Blue	Green	Purple	Teal
1	blue × green	0	4	4	0
2	green × purple	0	3	3	0
3	teal × blue	0	4	0	4
4	teal × purple	0	4	0	3
5	green × green	0	6	2	0

For each of the crosses shown above, deduce the parental genotypes where f represents the feather gene and the following superscript designations represent the different alleles:

b - blue; g - green; p - purple; t - teal

*Assume homozygosity unless otherwise indicated.

Problem 5

A geneticist discovered two pure breeding lines of ducks. One line had white eyes and quacked with a "quack-quack." The other line had orange eyes and had a deeper quack, "rock-rock." In order to determine the mode of inheritance of these characteristics, she mated the two types of ducks and found that all F_1 ducks had yellow eyes, and uttered "quack-quack." When the F_1 ducks were interbred, the F_2 ducks were found in the following ratios:

24	yellow-eyed, quack-quack
12	white-eyed, quack-quack
12	orange-eyed, quack-quack
6	yellow-eyed, rock-rock
3	white-eyed, rock-rock
3	orange-eyed, rock-rock
2	yellow-eyed, squawk
1	white-eyed, squawk
1	orange-eyed, squawk

How many genes are involved in the inheritance of eye color?

P   orange-eyed × white-eyed

F_1 yellow-eyed

F_2 2   yellow-eyed
            1   orange-eyed
            1   white-eyed

Use *A1/A2, 1/ B B2*, etc. to designate eye color alleles, assign genotypes to the individuals in the cross described above.

How many genes are involved in the inheritance of quacking?

P		quack-quack × rock-rock
F_1		quack-quack
F_2	12	quack-quack
	3	rock-rock
	1	squawk

Use *B/b*, *C/c*, etc. to designate quacking phenotype, assign genotypes to the individuals in the cross described above.

Problem 6

Recall from Problem 4 that blood types are classified by their surface antigens. Type A blood has antigen A on its surface, type B has antigen B, type AB has both antigen A and antigen B, and type O has neither antigen on its surface.

Ellen and Carl have just had a baby boy. Ellen's blood type is B, Carl's is AB and the baby's is O. Having a good knowledge of blood typing, you realize that something is not quite right with this story. You happen to know that Ellen's parents' blood types are B and O. From this, what is Ellen's genotype?

What is Carl's genotype?

Does it appear that Ellen and Carl can have a child with O blood?

Ellen's ex-boyfriend, Mark, has type O blood. Could Mark be the baby's father?

Ellen insists that Carl is the father, so you look into their family history (you may notice that Ellen and Carl are cousins). The results are seen in the pedigree below. Which parents have phenotypes that are incompatible with the blood types of their children?

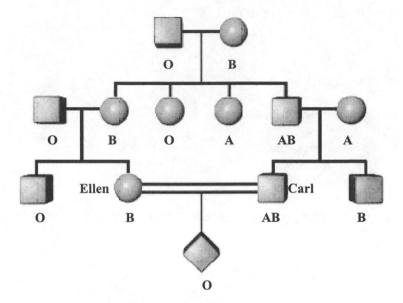

One of your friends suggests that this may be an example of lack of penetrance, i.e., the phenotype does not reflect the genotype. However, since Ellen and Carl share a grandfather who also has a suspicious O phenotype, you wonder if it's possible that they both inherited a recessive allele that is epistatic to the ABO blood antigens.

Upon further research you find there is such a rare recessive mutation, h, that is epistatic to the ABO system gene. Individuals who are homozygous for h cannot synthesize A or B antigens, so they have an O phenotype, referred to as the Bombay phenotype.

Assign genotypes for the H gene, assuming this is the reason for the unusual phenotypes (use H/h). With this information, determine the genotype for the ABO locus for individual I-1.

Does it now appear that Carl could be the father?

Does this information conclusively prove that Carl is the father?

In the hope of solving the paternity issue, you decide to run a Southern blot, probing for the alleles of the H gene.

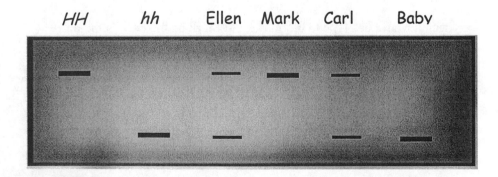

With this information, can you definitively say who is the baby's father?

Problem 7

Carolyn and Jeff are both cat lovers and neighbors in a local apartment complex. Recently, Carolyn's cat, Grace, escaped from her apartment and was seen mating with Jeff's ferocious feline, Chuck. Although Carolyn and Jeff were initially upset with the scandalous behavior of their pets, they felt reassured that Grace and Chuck would produce a beautiful and profitable litter of kittens. Weeks later, Carolyn and Jeff discovered the litter of kittens and were surprised to find one cat with curled ears.

As their close friend and local genetics expert, they turn to you to provide a genetic explanation for this strange occurrence. At this point, can you determine whether the curled-ear cat occurred as a result of a spontaneous dominant mutation or whether both parents were heterozygous for a recessive mutant allele?

In order to determine whether the curled-ear mutation is dominant or recessive to the normal condition, you mate the curled-ear cat with an unrelated normal cat. From this mating, *two curled-ear cats and two normal cats* are produced. Based on these results, is the curled-ear mutant allele dominant or recessive to the wild-type allele?

The allele responsible for curled ears in cats is located at the Ear locus. Using the allele symbols *c* and *c+*, assign genotypes to the cats in the cross described above. Be sure to follow the format shown below when assigning genotypes:

Homozygous individuals: *cc* or *c+c+*
Heterozygous individuals: *cc+* or *c+c*

What ratio of curled-ear cats to normal cats do you expect from the mating of two heterozygous curled-ear cats?

After performing the mating between two heterozygous curled-ear cats, you wait several weeks and examine the litter of kittens. The mating produced four curled-ear cats and two normal cats. From these results, can you be certain about the mode of inheritance of curled ears?

What results do you expect for a mating between a homozygous curled-ear cat and a normal cat?

What results do you expect for a mating between a heterozygous curled-ear cat and a normal cat?

You decide to perform 170 matings between curled-eared cats to determine whether curled ears are inherited in a simple Mendelian manner. You obtain 1020 kittens from your 170 matings between heterozygous curled-ear cats. If curled ears are inherited according to our predicted 3 to 1 ratio, how many curled-ear cats do you expect to see out of the 1020 kittens?

You only observe 677 curled-ear cats and 343 normal cats from the 170 matings. After re-counting the kittens several times, you are certain that the number of curled-ear cats is accurate. What is our observed ratio of curled-ear cats to normal-ear cats?

What is the most likely genetic explanation for the results we have gathered?

Linkage

....

Goals for Linkage Analysis:
1. Recognize the difference between linkage and independent assortment in dihybrid crosses.
2. Relate parental chromosome input to linkage phase to calculate recombination frequencies.
3. Be able to make maps from recombination frequencies.
4. Use three-factor crosses to distinguish order and distances between genes.

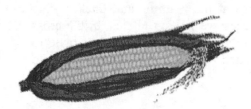

Problem 1

Consider two corn mutants: dwarf and glossy. Dwarf (gene symbol d) is a recessive trait characterized by short, compact plants. Glossy (gene symbol gl) is also recessive and is characterized by a bright leaf surface. F_1 progeny from a cross of pure breeding parents were back-crossed to the recessive parent. The results of this cross are shown below.

F_2:

286	wild type
89	glossy
97	dwarf
277	glossy dwarf

Since the ratio of phenotypic classes is not the expected 1:1:1:1 for a test cross, what concept best explains the observed phenotypic ratio?

What phenomenon explains the four phenotypic classes of unequal ratios instead of two phenotypic classes as predicted by the tight linkage?

Complete the genotype of each F_2 progeny class. Which phenotypic classes carry the recombinant genotypes?

What approximate percentage of progeny is descended from distinguishable recombinant gametes?

What is the map distance between dwarf and glossy determined from these numbers?

Using the known map distance between glossy and dwarf determined above, what is the expected number of each F_2 phenotypic class out of a total of 1000 F_2 progeny?

Problem 2

In *Drosophila*, the forked phenotype is characterized by short bristles with split ends. The scalloped phenotype is characterized by scalloped wings at the margins and thicker wing veins. Both genes are X-linked and are marked with recessive mutant alleles. In the cross below, an F_1 is generated by mating a wild-type female with a scalloped, forked male to produce all wild-type progeny. The females are then mated to their fathers (a test cross) to generate an F_2.

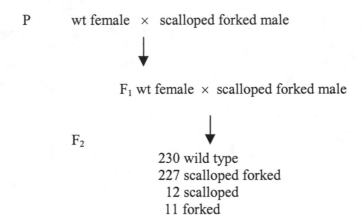

P   wt female × scalloped forked male

F_1 wt female × scalloped forked male

F_2

230 wild type
227 scalloped forked
12 scalloped
11 forked

What concept best explains the <u>observed</u> phenotypic ratio of the F_2 progeny?

Which of the phenotypes belongs to F_2 progeny that that are descended from distinguishable crossover bearing (recombinant) gametes?

What approximate fraction of progeny is descended from distinguishable crossover bearing gametes?

What is the map distance between forked and scalloped?

Problem 3

The *Drosophila* forked phenotype is X-linked recessive (as we saw in Problem 2) and is characterized by short bristles with split ends. The miniature phenotype, also X-linked recessive, is characterized by reduced wing size.

A miniature female is crossed with a forked male. The F_1 progeny consists of wild-type females and miniature males. These are then mated to yield the following F_2:

510 miniature females	418 miniature males
490 wild-type females	412 forked males
	87 miniature forked males
	83 wild-type males

What is the approximate distance in map units (mu) between forked and miniature that you calculate from this cross?

When a wild-type female is crossed with a miniature forked male, and the wild-type F_1 females are testcrossed, the F_2 are as shown below:

84 miniature
86 forked
417 miniature forked
413 wild type

Is the distance you find in mu the same in this cross as the distance you found for the markers in trans?

In Problem 2 you found that the distance between the scalloped gene and the forked gene is 5 mu. In this problem you found that the distance between miniature and forked is 17 mu. Is this information sufficient to find the gene order of the three genes on the X chromosome?

When a wild-type female is crossed with a miniature scalloped male, all the progeny are wild type. A testcross of the heterozygous females yields the following F_2:

51 miniature
45 scalloped
354 miniature scalloped
350 wild type

What is the approximate map unit (mu) distance between miniature and scalloped that you calculated from this cross?

Now that you know the distances between each pair of genes, construct a map of the three genes.

Problem 4

In *Drosophila*, the spineless mutant is characterized by shortened bristles compared to the longer bristles of the wild type. The radius incomplete mutant is characterized by an incomplete wing vein pattern. Both genes are autosomal and are marked by recessive mutant alleles.

A wild-type female is mated with a spineless, incomplete male. Then, an F_1 wild-type female is crossed to the parental male to generate the following F_2:

> 449 wild type
> 452 spineless incomplete
> 52 spineless
> 48 incomplete

In this cross you did not obtain the expected 1:1:1:1 ratio expected for a test cross of unlinked genes. This implies that the genes are linked. Use these data to calculate the distance between spineless and incomplete.

A wild-type male is mated with a spineless, incomplete female. Then, the F_1 wild-type male is mated with the parental female. The following F_2 progeny are observed:

> 361 wild-type flies
> 357 spineless incomplete flies

Note that very different results are observed in this F_2 compared to the first cross. To help explain these unusual results, let's first assign genotypes to the F_1 male and tester female.

Write the genotypes of the F_1 flies in the cross above. Use the symbols *sp, sp+* for spineless and *ri* or *ri+* for radius incomplete. Indicate linkage by writing the two linked alleles on one side of a / (e.g., *a+b+/ab*).

After examination of these genotypes, does it make sense that these two genes are on the same chromosome?

Geneticists made a novel discovery when working with *Drosophila melanogaster*. Scientists found that male flies do <u>not</u> undergo recombination. Therefore, test cross of the F_1 male yields only two gamete types. The distance between genes cannot be measured using F_1 males. What gametes are produced in the F_1 male fly?

The consequence of no recombination in males is that crosses to map linked genes must use F_1 females as the dihybrid parent. This is true only in *Drosophila melanogaster*. In other species there is frequently a difference in recombination between the sexes, but it is not usually 0 in males.

Problem 5

In *Drosophila*, the X-linked genes singed (*sn*), characterized by bent bristles, miniature (*m*), reduced wing size, and tan (*t*), a tan body color, are marked with recessive alleles. In this problem you will use three-factor crosses to determine the gene order and distance between these markers. (Mating the F_1 flies is essentially a test cross, allowing one to analyze all of the F_2 flies directly).

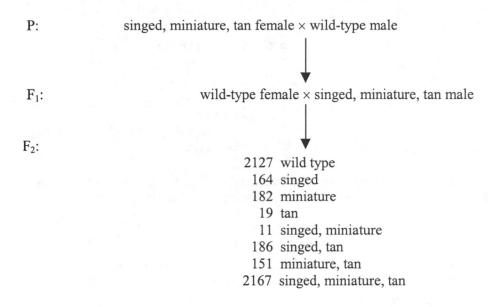

P: singed, miniature, tan female × wild-type male

F_1:   wild-type female × singed, miniature, tan male

F_2:

2127 wild type
164 singed
182 miniature
19 tan
11 singed, miniature
186 singed, tan
151 miniature, tan
2167 singed, miniature, tan

How many gamete types do you expect the F_1 female to produce?

Which two classes of gamete types are the produced most often by the F_1 female? (This is reflected in the phenotypic classes of the progeny.) These are the parental types.

Which of the reciprocal gamete types are the produced least often in the F_1 female? (This is reflected in the phenotypic classes of the progeny.) These are the double crossover types.

Compare the double crossover gametes with the parental gametes to determine which gene is in the middle. Write the correct gene order below.

Using your data, determine the distance between the singed gene and the tan gene, and the distance between the tan gene and the miniature gene.

Problem 6

Chromosome 13 of the mouse carries the locus for flexed tail with the alleles f and $f+$ and the locus for extra toes with the alleles Et and $Et+$. The flexed tail mutant is a recessive trait while the extra toes mutant is dominant. The next exercises will use 24 map units as the known distance between Et and f to predict the expected number of F_2 progeny from crosses.

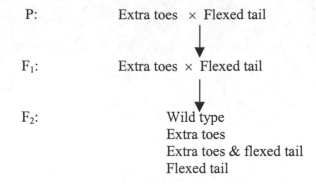

P: Extra toes × Flexed tail

F₁: Extra toes × Flexed tail

F₂: Wild type
                        Extra toes
                        Extra toes & flexed tail
                        Flexed tail

Find the genotype of each mouse above. Both parental mice are from pure breeding stocks. Use Et and $Et+$ for the extra toes gene, and f or $f+$ for the flexed tail gene.

Which F_2 progeny represent recombinants?

The distance between Et and f is 24 mu. From a total of a 1000 F_2 mice obtained from crosses such as this, how many are expected for each phenotypic class? (Start with the number of wild-type progeny.)

                   Wild type
                   Extra toes
                   Extra toes & flexed tail
                   Flexed tail

Let's now try to find the expected progeny for a three point cross. Chromosome 13 of mouse also carries the locus for satin fur texture with the alleles sa and $sa+$. The satin fur mutant is a recessive trait. Use the following map unit distances to map sa: $sa – Et$: 4 mu, $sa – f$: 20 mu. Draw a map of the three markers.

Use this map to determine the number of predicted progeny of each phenotypic class shown below with parents of genotypes $Et\ f/Et\ f$ and sa/sa. The F_2 progeny classes from these crosses are as follows:

                   Wild type
                   Flexed tail, satin with extra toes
                   Flexed tail with extra toes
                   Satin with extra toes
                   Flexed tail, satin
                   Extra toes
                   Flexed tail
                   Satin

Bacterial Genetics

Goals for Bacterial Genetics:
1. Be familiar with the mechanism of conjugation and transduction as methods of gene transfer.
2. Determine genotype from ability to grow on different media.
3. Determine type of media needed for selection of exconjugants.
4. Be familiar with replica plating and interpretation of resulting data.
5. Map genes relative to other genes by interrupted mating, natural gradient of transfer, and variations of these techniques.

Conjugation
Problem 1

Various strains of bacteria were incubated on plates containing minimal media plus several amino acids. From the pattern of growth, answer the following questions regarding the genotype of each strain.

What are the genotypes of colonies 1–4 with respect to *arg*, *met*, *asp*, and *ile*?

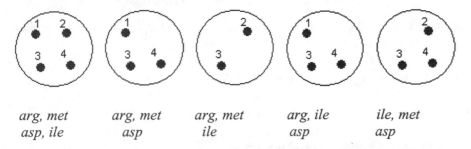

| arg, met | arg, met | arg, met | arg, ile | ile, met |
| asp, ile | asp | ile | asp | asp |

The following plates show bacterial strains that have been replica plated on minimal media plus the indicated mixture of amino acids and antibiotics.

Determine the genotypes of colonies 1–4 with respect to *arg*, *met*, *pen*, and *str*.

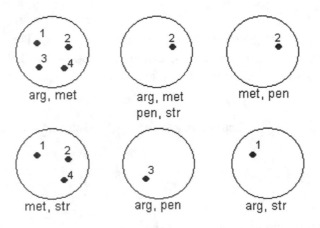

arg, met

arg, met
pen, str

met, pen

met, str

arg, pen

arg, str

The plates shown below contain different sugars as carbon sources. Cells were first plated

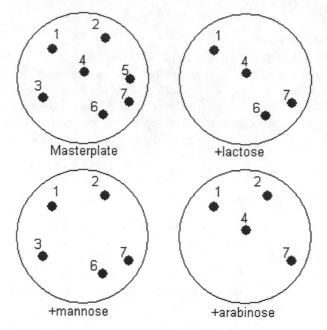

on minimal media with glucose as the carbon source (Masterplate) and then replica plated onto plates with lactose, mannose, or arabinose (no glucose).

Which colony has the genotype *lac– man+ ara+*?

Which colony has the genotype *lac+ man– ara+*?

Which colony has the genotype *lac+ man+ ara–*?

Which colony has the genotype *lac– man+ ara–*?

Which colony has the genotype *lac– man– ara–*?

Problem 2

Lederberg and Tatum made crosses by mixing pairs of strains. They used a C strain which was *arg–met+* and an E strain which was *arg+met–*. However, they did not know whether each of these strains was F–, F+ or Hfr. An example of the type of crosses they performed is shown below.

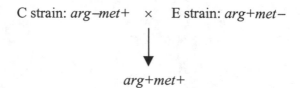

C strain: *arg–met+*   ×   E strain: *arg+met–*

arg+met+

What type of cross would give you many exconjugants?

Which of the following crosses will give you only a few exconjugants?

What kind of crosses will give no *arg+ met+* exconjugants? (select from the choices below)

F+ × F+   **F− × Hfr**   **F+ × Hfr**

F− × F−   **F+ × F−**   **Hfr × Hfr**

The table shown here contains the results of all possible crosses (0 = none; F = few; M = many). From the questions you just answered, you should now be able to identify which strains are F+, F−, and Hfr.

	E1	E2	E3	E4	E5
C1	F	0	M	F	0
C2	0	M	0	0	M
C3	0	F	0	0	F
C4	F	0	M	F	0
C5	0	F	0	0	F

Bonus Question

What we are looking for in our exconjugants are the recombinants of the E strains, *arg+met−*, with the C strains, *arg−met+*. What type of media should be used to select for the prototrophic exconjugants?

Problem 3

From the time that Hfr and F− cells are combined, the number of markers transferred for each mating pair depends on how long the two cells stay conjoined. The mating can be interrupted at specified times after the start of mating by vigorous shaking, frequently accomplished by blending. Without such mixing the disruption of the mating pairs occurs naturally generating a gradient of transfer.

In the following experiment, a prototrophic Hfr strain (streptomycin sensitive) is mated to a streptomycin resistant F− strain auxotrophic for methionine, leucine, and cysteine. A pipette is used to distribute cells to each of the plates shown below, each containing minimal media with streptomycin, methionine, and cysteine added in order to select only for leucine prototrophs.

Aliquots of culture A (Hfr cells only), culture B (Hfr cells + F− cells), and culture C (F− cells only) are added to the three plates containing streptomycin, methionine, and cysteine. The plates are then incubated. The results are shown in the following illustration. Genotypes: (Hfr: *str-s, met+, leu+, cys+*) (F−: *str-r, met−, leu−, cys−*).

A: *str, met, cys*   **B: *str, met cys***   **C: *str, met, cys***
No colonies   **375 colonies**   **No colonies**

Why were no colonies observed after plating the Hfr strain on plate A above?

Why were no colonies observed after plating the F- strain on plate C above?

Why were cells able to grow on plate B?

The mating mixture produced 375 exconjugant colonies that replaced the *leu–* allele of the F– strain with the *leu+* allele of the Hfr strain. Additional markers may have also been transferred and could be checked by replica plating.

To measure the time of entry of each of the markers, matings between the Hfr and F– strains can be interrupted at various time points and plated on selective media. Below is a graph of colonies versus time.

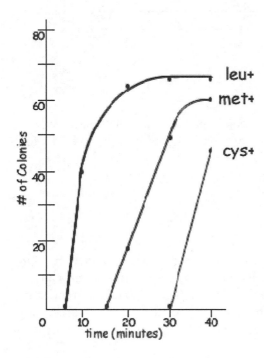

Draw a map below to indicate the order of the markers in the Hfr strain with respect to the origin.

Problem 4

An Hfr strain with genotype *met– leu+ his+ trp+* that transfers the *met* gene very late was mated with an F– *met+ leu– his– trp–* strain. After mating for 30 minutes, cells were plated on minimal media (MM) with the added nutrients listed below each plate. The number of colonies that grew on each plate is indicated.

Plate	1	2	3
Supplements	his, trp	his, leu	leu, trp
Colonies	250	50	500

What is the purpose of the methionine marker in this cross?

What markers are selected for on Plate 1?

What markers were selected on Plate 2?

What markers were selected on Plate 3?

Those markers closest to the origin are transferred first and yield the highest number of recombinant colonies. Based on the number of colonies on each plate above, determine the order of the markers in the Hfr strain with respect to the origin.

Problem 5

A small bacterial genome was mapped using three different Hfr strains and an F– *str'* *ala–* *ade–* *bio–* *his–* *ile–* *val–*. Each of the Hfr strains was *str^s* and contained the wild-type alleles for all the markers. The matings were interrupted after 30 minutes and the exconjugants were selected on the following plates (the indicated nutrient is left out of the otherwise complete media + streptomycin).

Hfr: *str^s ala+ ade+ bio+ his+ ile+ val+*   ×   F–: *str' ala– ade– bio– his– ile– val–*

Mating	Missing Nutrient					
	Ala	**Ade**	**Bio**	**His**	**Ile**	**Val**
HfrA × F-	300	0	0	0	900	750
HfrB × F-	0	400	0	0	700	875
HfrC × F-	200	0	956	724	0	0

In the table above, some of the plates have no colonies. How can this be explained?

Consider the data from the Hfr A × F– cross. Which of the markers can be ordered using this cross?

What is the first marker to be transferred by Hfr A?

Order the three markers starting with *ile*.

What is the order of the markers indicated by the results of the Hfr B × F– cross?

What is the order of the markers indicated by the results of the third cross?

In every problem so far we have represented the bacterial chromosome as a circle. The original recognition of this fact comes from comparing the maps of different Hfr strains. As you can see from your results, the maps can only be reconciled as permutations from a common circular map, with the different Hfr strains having different origin points and directions of transfer. Use this information to construct a circular map.

Transduction
Problem 1

A generalized transducing phage is grown on a prototrophic strain and then used to transduce a recipient that is *arg– leu– gln– gua– thr–*. In this experiment *arg+* transductants will be selected by plating on media lacking arginine but containing all other supplements required by the recipient strain. This will be the master plate.

DONOR: *arg+ leu+ gln+ gua+ thr+*
RECIPIENT: *arg– leu– gln– gua– thr–*

MASTERPLATE: leu, gln, gua, thr (supplements)

Fifty-nine colonies grew on this master plate. To test for co-transduction of the unselected markers, the colonies were replica plated to different media, omitting one supplement at a time. The results are shown in the following table.

Supplements to Minimal Media					
Plate	Leucine	Guanine	Glutamine	Threonine	Colonies
1	–	+	+	+	0
2	+	–	+	+	0
3	+	+	–	+	24
4	+	+	+	–	0

What does the absence of growth indicate on Plate 1?

What does the presence of growth indicate on Plate 3?

What is the co-transduction frequency (in percent) of *arg* and *gln*?

To develop a genetic map, let us now repeat the transduction experiment and make a new master by plating on media lacking leucine but containing all other supplements required by the recipient strain. This is our second master plate. Seventy-three colonies grew on this plate and were replica plated onto various plates containing the media indicated in the table below.

Supplements to Minimal Media					
Plate	Arginine	Guanine	Glutamine	Threonine	Colonies
1	–	+	+	+	0
2	+	–	+	+	22
3	+	+	–	+	0
4	+	+	+	–	45

What is the co-transduction frequency (in percent) of *leu* and *gua*?

What is the co-transduction frequency (in percent) of *leu* and *gln*?

To finish our experiment, we will do one more transduction. In this third master plate our media lacks guanine but contains all other supplements required by the recipient strain. Sixty-five colonies grow on this plate which is replica plated onto various plates containing the media indicated in the table below.

Supplements to Minimal Media					
Plate	Arginine	Leucine	Glutamine	Threonine	Colonies
1	–	+	+	+	0
2	+	–	+	+	20
3	+	+	–	+	13
4	+	+	+	–	0

What is the co-transduction frequency of *gua* and *gln*?

Using these co-transduction frequencies you have determined, determine the order of the three markers: *arg, gua,* and *gln*.

Problem 2

In a transduction experiment, the donor strain is *kan^r lys+ arg+* and the recipient strain is *kan^s lys– arg–*. Transductants are plated on MM (minimal media) supplemented with kan (kanamycin), lys (lysine), and arg (arginine) and then replica plated on the plates shown below.

Supplements to Minimal Media				
Plate	Kan	Lys	Arg	# of colonies
Master	+	+	+	500
Replica 1	+	–	–	20
Replica 2	+	+	–	21
Replica 3	+	–	+	200

In the table below, fill in the number of colonies for each of the four genotypes.

Genotype	# of colonies
kanr arg+ lys+	
kanr arg+ lys–	
kanr arg– lys+	
kanr arg– lys–	

What is the co-transduction frequency of *kan* and *arg*?

What is the co-transduction frequency of *kan* and *lys*?

Problem 3
This problem is a simulation and can only be done on the *Interactive Genetics* CD-ROM.

Problem 4

Frequently, mutations are so close together that it's impossible to determine their order with respect to a nearby marker by simply using conjugation or transduction.

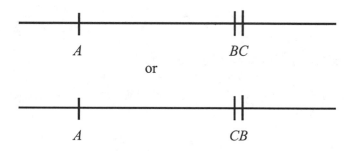

In the illustration above it is difficult to determine if *B* is closer to *A* or *C* is closer to *A*. The co-transduction frequency between *A* and *B* would be about the same as the co-transduction frequency between *A* and C. However, reciprocal three-factor transductions can be used to determine the order of these markers.

Unlike in plants and animals, reciprocal transductions in bacteria do not refer to switching markers between sexes. Instead, in bacteria we refer to switching the markers between the donor and recipient. That is, in Transduction I the donor may be *B+C−* and the recipient *B−C+*, while in Transduction II (the reciprocal) the donor may be *B−C+* and the recipient *B+C−*. Having the markers in trans is essential for this experiment since it forces a crossover between the two tightly linked markers (*B* and *C*) to generate a prototroph. The outside marker will be *A+* in the donor and *A−* in the recipient in all transductions.

In the two reciprocal transductions shown below the order is assumed to be *ABC*. Draw an X in the appropriate places to generate prototrophic (*A+B+C+*) transductants for each experiment.

Transduction I: Donor: *A+B+C−*   Recipient: *A−B−C+*

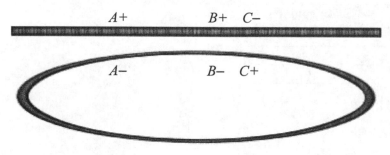

Transduction II: Donor: $A+B-C+$ Recipient: $A-B+C-$

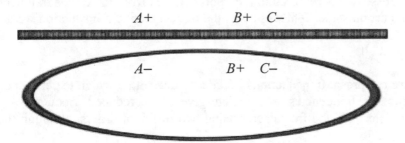

Notice that in Transduction I, prototrophs were generated with only two crossovers, whereas in Transduction II, four crossovers were required to generate $A+B+C+$. This is true only when the order is ABC.

Now we will see what happens if the order is ACB. Again, draw X's in the appropriate places to generate prototrophic ($A+B+C+$) transductants.

Transduction I: Donor: $A+B+C-$ Recipient: $A-B-C+$

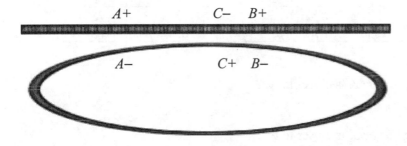

Transduction II: Donor: $A+B-C+$ Recipient: $A-B+C-$

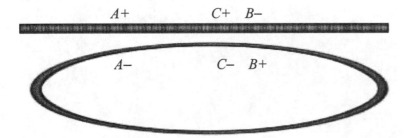

Notice that Transduction I prototrophs were generated with four crossovers, whereas in Transduction II only two crossovers were required to generate $A+B+C+$.

A minimum of two crossovers is required to incorporate genes from the donor fragment into the circular recipient chromosome. This results in the frequencies of transduction we usually observe. However, a situation that requires four crossovers will yield many fewer transductants or none at all.

You need to compare reciprocal transductions in order to determine which experiment gives you normal transduction frequencies and which gives you reduced frequencies. This example asks you to determine the frequencies using two different orders, although there is only one order.

Reciprocal transductions were used to order three mutations (*trp1, trp2,* and *trp3*) required for metabolism of tryptophan with respect to a nearby marker tyrosine (*tyr+*). For each pair of mutants (*trp1* and *trp2; trp2* and *trp3; trp1* and *trp3*) a pair of reciprocal transductions will be made to order them with respect to *tyr*.

	Donor	Recipient
Transduction I	*tyr+ trpx– trpy+*	*tyr– trpx+ trpy–*
Transduction II	*tyr+ trpx+ trpy–*	*tyr– trpx– trpy+*

Experiment	x	y	I	II
1	*trp1*	*trp2*	800	5
2	*trp2*	*trp3*	3	750
3	*trp1*	*trp3*	659	8

Which mutation, *trp1* or *trp2*, is closer to *tyr*?

Which mutation, *trp1* or *trp3*, is closer to *tyr*?

Which mutation, *trp2* or *trp3*, is closer to *tyr*?

Draw a map below to indicate the order of the three mutations with respect to *tyr*.

Biochemical Genetics

Goals for Biochemical Genetics:
1. Determine the number of genes coding for steps in a biosynthetic pathway. Use auxotrophic mutant strains to determine the number of complementation groups for a pathway.
2. Determine a biosynthetic pathway from nutritional studies that supplement with compounds that may be intermediates in the pathway.
3. Determine whether any of the genes are linked to each other.
4. Distinguish the metabolism of amino acids between anabolic (biosynthetic) and catabolic (degradative) reactions. Most human biochemical disorders show up from the loss of a catabolic enzyme. Auxotrophic mutant strains of microorganisms result from the loss of an anabolic enzyme.

Problem 1

Six *Neurospora* mutants were isolated that require vitamin B1 to grow. To determine whether the six mutants have mutations in different genes, complementation studies were performed. Heterokaryons were formed and tested for growth on minimal media.

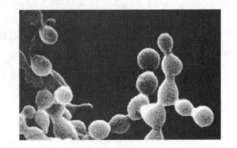

The following table shows the experimental results; + indicates heterokaryon growth and − indicates no heterokaryon growth in the absence of vitamin B1.

	A	B	C	D	E	F
A	−	+	+	−	+	+
B	+	−	+	+	−	−
C	+	+	−	+	+	+
D	−	+	+	−	+	+
E	+	−	+	+	−	−
F	+	−	+	+	−	−

Which mutants belong to the same complementation group as mutant A?

Which mutants belong to the same complementation group as mutant B?

Which mutants belong to the same complementation group as mutant C?

Which mutants belong to the same complementation group as mutant D?

How many complementation groups are there?

Would a heterokaryon formed from mutant A and the double mutant C,E grow on minimal media? The double mutant strain contains two gene mutations: the gene mutation in strain C and the gene mutation in strain E.

Would a heterokaryon formed from the double mutants A,B and C,F grow on minimal media?

Would a heterokaryon formed between the three double mutants A,E and B,C and C,D grow on minimal media?

Problem 2

You are a graduate student working in a *Neurospora* lab and have recently isolated five *Neurospora* methionine auxotrophs that each contain a single gene mutation. You set out to determine if any of the mutated genes in the *Neurospora* strains are linked. You decide to do this by crossing mutant strains and examining the phenotypes of the progeny.

You form a transient diploid by mating mutant strains (represented by the cross *ab+ × a+b*). The diploids then undergo meiosis immediately by sporulation. A diploid cell with a crossover in the interval between the linked *a* and *b* genes (in this example) will yield four genetically distinct spores. Three of them are still auxotrophs and only the *a+b+* spore is phenotypically distinct as a prototroph. The frequency of prototrophs in a large collection of spores can be used to determine the map distance between two mutants.

The results of your crosses with the five *Neurospora* auxotrophic strains (A–E) in all combinations are shown below. For each cross, 1000 ascospores were plated on minimal media. The table shows the number of methionine prototrophs that you recovered from each cross.

	A	B	C	D	E
A	0	150	250	215	40
B		0	250	65	110
C			0	250	250
D				0	175
E					0

Since strains A–E are methionine auxotrophs, the prototrophic ascospores must have been produced from independent assortment or recombination between two mutated genes during meiosis. How many prototrophic ascospores would you expect from a cross if the recombinant progeny were produced by independent assortment?

For each of the linked genes in the table above, determine the map distance between them. Then, combine the information from each of the two factor crosses to assemble a map of the entire linkage group.

Problem 3

Wild-type *Neurospora* is orange when exposed to light during growth. You have isolated three albino mutants that are completely white even when grown in the presence of light. Using heterokaryon complementation studies, you find that these three strains have mutations in different genes, which you name *al-1*, *al-2*, and *al-3*.

You suspect that the albino mutants are white because they are unable to make the carotenoid pigment (known to cause the orange color). To test your idea, you seed the albino mutants on media supplemented with carotenoid pigment.

You are thrilled to find that supplementation with carotenoid pigment results in orange colored hyphae for all three albino mutants! Assuming carotenoid synthesis is affected in these mutant strains, you set out to determine which step in the biosynthetic pathway is blocked by each of the mutations.

Luckily, three precursors in the carotenoid pigment biosynthetic pathway had previously been discovered, although their order was unknown. To determine the order in which the precursors are converted into the carotenoid pigment, you grow the albino mutants on media supplemented with each of the three precursors.

You find that supplementing media with these different precursors restores wild-type color in some of the albino mutant. You compile your results into a simple table, where + indicates wild-type color and − indicates white color.

	Carotenoid pigment	GGPP	Phytoene	PPP
al-1	+	−	−	−
al-2	+	−	+	+
al-3	+	−	+	−

Use the data from your experiments to determine the order of the precursors as they appear in the biosynthetic pathway and which step in the biosynthetic pathway is blocked by each mutant.

Problem 4

Saccharomyces cerevisiae has played a fundamental role in human history and culture. For centuries, yeast has been used by humans for the rising of bread and for the fermentation of wines and beers. Today yeast is used as a model organism for both genetic and cellular biology studies. Yeast are eukaryotes that grow simply as either single cells or colonies.

Saccharomyces cerevisiae can exist as either a haploid or a diploid. Both haploids and diploids are able to grow and divide by mitosis, through a process called budding. This differs from *Neurospora* where the diploid zygote is transient and quickly undergoes meiosis to form ascospores. In the haploid state there are two mating types, a and α. When an a cell and an α cell are brought together, they fuse to form a diploid cell. First the cellular membranes fuse, then the nuclei (yeast do not normally form stable heterokaryons). The diploid cell can grow and divide indefinitely. However, when nutrients are depleted or when environmental conditions become unfavorable, the diploid cells will undergo meiosis to form four haploid spores (a process called sporulation). The spores can be separated manually and tested directly for phenotype. Since they are haploid, their phenotype directly reflects their genotype.

Ten mutant yeast strains have been isolated that cannot grow on medium with galactose as the sole carbon source (galactose medium). The ten strains are named A-J. Each mutant was separately crossed to the others to form a set of diploid strains. The ability of the diploids to grow on galactose medium was used as a measure of complementation. The results of this experiment are given in the chart below.

Gal Mutants

	A	B	C	D	E	F	G	H	I	J
A	–	+	+	+	+	–	–	+	–	+
B		–	+	+	+	+	+	–	+	+
C			–	+	+	+	+	+	+	+
D				–	–	+	+	+	+	–
E					–	+	+	+	+	–
F						–	–	+	–	+
G							–	+	–	+
H								–	+	+
I									–	+
J										–

Are mutants A and B in the same complementation group?

Are mutants A and F in the same complementation group?

How many complementation groups are present?

To determine whether the four genes identified were linked to one another, each diploid strain was sporulated and 1000 random spores were analyzed for growth on galactose medium. The number of spores that were able to grow on galactose medium is indicated in the table below.

	A	B	C	D	E	F	G	H	I	J
A	0	250	0	0	0	0	0	250	0	0
B		0	250	250	250	250	250	0	250	250
C			0	0	0	0	0	250	0	0
D				0	0	0	0	250	0	0
E					0	0	0	250	0	0
F						0	0	250	0	0
G							0	250	0	0
H								0	250	250
I									0	0
J										0

Since we know certain strains contain mutations in the same gene, we can simplify the chart above to reflect the same data by grouping together mutants from the same complementation group. We will arbitrarily give each complementation group a *gal* gene designation as follows: Mutants A, F, G, I—*gal1*; Mutants B, H—*gal4*; Mutant C—*gal7*; Mutants D, E, J—*gal10* (these designations are given based on actual *gal* genes in yeast).

	gal1	*gal4*	*gal7*	*gal10*
gal1	0	250	0	0
gal4		0	250	250
gal7			0	0
gal10				0

Is *gal1* linked to *gal4*?

gal1 shows no wild-type recombinants with either *gal7* or *gal10*. How can you explain this observation?

Analysis of 100,000 haploid spores produced from each of the crosses above were plated on galactose medium. Use this data to determine the map distance between *gal1* and *gal7*, *gal1* and *gal10*, and *gal7* and *gal10*.

	gal1	*gal4*	*gal7*	*gal10*
gal1	0	25000	35	16
gal4		0	25000	25000
gal7			0	18
gal10				0

Problem 5

This problem is a simulation and can only be done on the *Interactive Genetics* CD-ROM.

Population Genetics

Goals for Population Genetics:
1. Be able to relate allele frequencies at a gene to population homozygote and heterozygote frequencies.
2. Be able to use Chi-square test to determine if a population is in Hardy-Weinberg equilibrium.
3. Combine family studies with population frequencies to predict the chance a child will be homozygous for a disease allele.

Problem 1

Do you know your genotype with respect to your ABO or MN blood types? Will you consider these genotypes when you decide to marry? Most people are unaware of the alleles they carry for the majority of their genes. How do you study the genetics of animals in a natural environment, where family units are usually impossible to discern (e.g., fish). Because of the lack of pedigree data, individuals are regarded only as samples from the larger population.

A population, consisting of interbreeding individuals in a prescribed geographical area, contains a reservoir of all the gene copies (alleles) that will give rise to the individuals in the next generation. In these and similar examples, the population's allele frequencies are used to predict the genotype frequencies of the individuals.

The reservoir of alleles for a single gene is referred to as the gene pool for that gene. The genotypes of individuals can be considered a random (unbiased) sampling from the gene pool, with a gamete representing a single sample from the gene pool. In the following sets of questions we are going to use a bowl of ping-pong balls (whose different colors represent different alleles) to simulate random sampling.

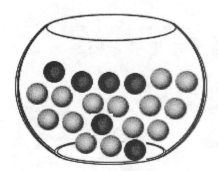

To represent diploid individuals subsequent questions will use pairs of samplings. What is the frequency of black ping-pong balls in the adjacent bowl?

Note that the probability of randomly obtaining a black ball, represented by p, is the frequency of black balls in the bowl.

What is the frequency of the white ping-pong balls in this bowl?

The frequency of white balls, represented by q, is the probability of randomly obtaining a white ball. Here black and white are the only colors of Ping-Pong balls in this bowl, and $p + q = 1$. To represent diploid individuals subsequent questions will use pairs of samplings.

What is the probability of picking two black balls (homozygous black)?

What is the probability of picking two white balls (homozygous)?

What is the probability of picking one white ball and one black ball (heterozygous) in either order?

As you have probably noticed, the sum of the probabilities found equals 1, that is $p^2 + 2pq + q^2 = 1$. This equation, which is derived from $(p + q)^2 = 1$ (when there are two alleles for a certain trait), describes the expected genotype frequencies of a population in Hardy-Weinberg equilibrium.

Problem 2

Consider this hypothetical population of 25 individuals of the FISH species. The sum total of all alleles present in the population is referred to as the population's gene pool. We will examine only one gene locus with two possible alleles: F for green color, and f for white color. In this population, the solid green fish are homozygous for F, the solid white fish are homozygous for f, and the spotted fish are heterozygous. Answer the following questions using the observed phenotypes.

Using the fish above, calculate p, the frequency of the green allele.

What is q, the frequency of the white allele? Recall, in the population above, ten fish are FF, five fish are ff, and ten fish are Ff.

To determine if this population is in equilibrium you need to determine the expected numbers using the Hardy-Weinberg equilibrium equation. To begin with, what is the expected frequency of FF individuals?

What is the expected <u>number</u> of FF individuals in this population?

What is the expected number of ff fish? What is the expected number of Ff individuals?

To find whether or not the population depicted above is in HWE (Hardy-Weinberg equilibrium), we will perform a chi-square (χ^2) test, using expected and observed values of each genotype. The χ^2 test is used to determine whether deviation from expected values are due to chance alone. If the deviations are too large we conclude that something besides chance is involved and the population is not in HWE. What is the value of χ^2 for this example? This χ^2 value can be converted into a probability value. To do this, we need the number of degrees of freedom (df) for the particular χ^2 test. The df equals the number of variables - 1. Although there are three genotypic classes, their numbers are determined by only two variables (p and q). Therefore, the degrees of freedom is 1. Using a χ^2 distribution chart, determine the probability (the p value) of obtaining these deviations due to chance alone. From your χ^2 test, is the population depicted above in Hardy-Weinberg equilibrium (HWE)?

Problem 3

Men are from Mars. Women are from Venus. When Mars and Venus collide, a new population arises. The allele frequencies on Mars and Venus differ. In the following problem, two alleles of a gene are represented by gray and blue ping-pong balls.

Mars   Venus

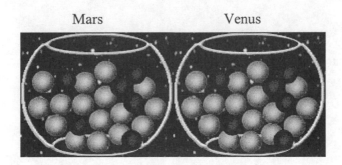

Predict the genotype frequency for homozygous gray in the new population.

What proportion of homozygous blue individuals do you expect in the new population?

What is the proportion of heterozygotes you expect the new population to have?

If the new population breeds only among themselves, will the next generation be in Hardy-Weinberg equilibrium?

Problem 4

The ability to taste the chemical phenylthiocarbamide (PTC) is an autosomal dominant phenotype. Tasters can detect it as extremely bitter, while nontasters cannot detect it at all. The genotype frequency of the homozygous recessive individuals (nontasters) in the U.S. population is 0.35, or 35%. Note that this population is in Hardy-Weinberg equilibrium (HWE) for this trait. What is the allele frequency for the nontaster allele?

q is commonly used to represent the allele frequency of the recessive (nontaster) allele. q^2, therefore, represents the genotype frequency of the homozygous recessive, nontaster individuals with a value of 0.35. What is the allele frequency for the dominant taster allele?

p is commonly used to represent the allele frequency of the dominant (taster) allele. After finding allele frequencies, we will use these values to find genotype frequencies. Answer the following questions using p and q. What is the frequency of heterozygotes in the general population?

What is the frequency of heterozygotes among the PTC taster population only?

A PTC taster man whose mother is a taster while his father is not, marries a taster woman. Recall that PTC tasting is an autosomal dominant trait. What is the probability that their child will be a nontaster?

Problem 5

Red green colorblindness is a recessive sex-linked trait. In a given population that exists in HWE, one in every eight males is colorblind. Using this information answer the following questions. What is the allele frequency for the red green colorblindness allele? (Or, what is q?)

What proportion of all women are colorblind?

In what proportion of marriages will all the males be colorblind but none of the females?

In what proportion of the marriages will all the males be normal and the females be heterozygous?

Molecular Markers

Goals for Molecular Markers:
1. Recognize the kinds of variation in DNA sequences between homologous chromosomes that can be used as codominant alleles, including restriction fragment site polymorphisms and variations in the number of repeat sequences.
2. Understand the techniques (Southern Analysis and PCR) used to detect these variations within a defined chromosome region.
3. Understand how molecular markers are used for linkage studies to locate a disease gene and to start the positional cloning of the gene.
4. Use molecular markers to determine the genotype of an individual for disease prediction, for relationship studies between individuals, and for forensic identification of unknown individuals.

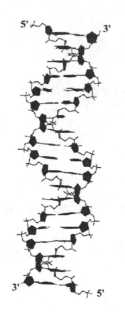

Problem 1

Restriction fragment length polymorphisms are identified by screening DNA isolated from members of families, using an array of different restriction enzymes. Random human genomic clones are then used as probes in a Southern blot. In the following problems you must identify the probe and restriction enzyme that gives rise to a polymorphism and then determine if the RFLP is linked to the inherited trait.

The DNA sample from each person is split into six tubes and digested with the following enzymes:

Apa I, Bam HI, Eco RI, Hind III, Sal I, Xma I

Gel electrophoresis separates the DNA fragments according to size. Denaturation of the DNA and transfer to nitrocellulose allows specific bands to be detected by hybridization with a radioactive probe (Southern blot). Choose the probe and restriction enzyme below that has an RFLP linked to the disease gene indicated by the pedigree.

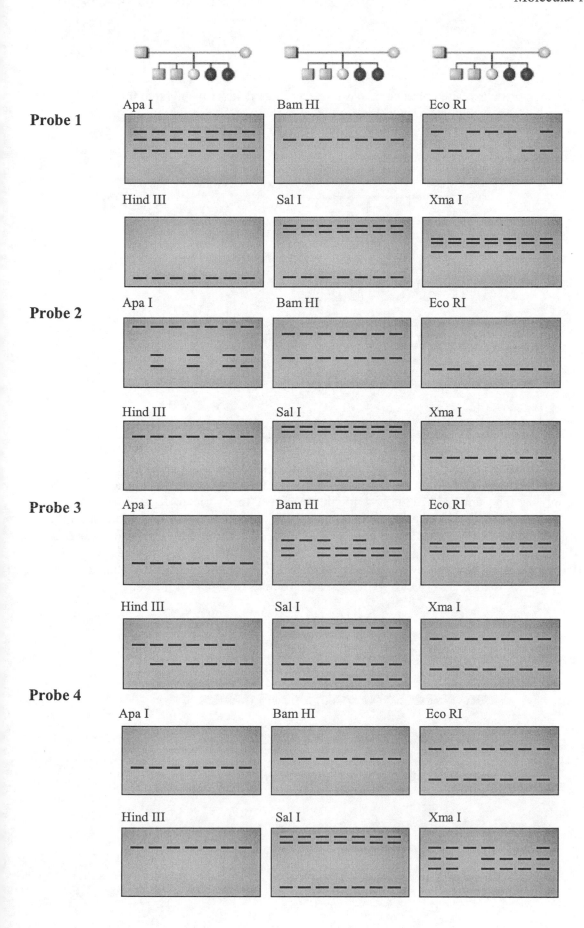

Probe 1

Apa I · Bam HI · Eco RI

Hind III · Sal I · Xma I

Probe 2

Apa I · Bam HI · Eco RI

Hind III · Sal I · Xma I

Probe 3

Apa I · Bam HI · Eco RI

Hind III · Sal I · Xma I

Probe 4

Apa I · Bam HI · Eco RI

Hind III · Sal I · Xma I

Problem 2

Part a

In the pedigree below, the presence of the rare recessive disease phenylketonuria (PKU) is indicated by filled symbols. Using the *A/a* above, assign genotypes based on the pedigree analysis.

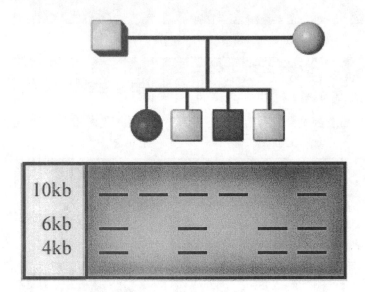

DNA was obtained from each of the individuals in the pedigree above, digested with a restriction enzyme and probed with a fragment known to hybridize to a linked RFLP. Analysis of the Southern blot above in conjunction with the pedigree allows determination of the linkage between the *A/a* gene and the RFLP shown. Focus first on the affected children. Use this information to determine the coupling (linkage) in each of the parents and then in the unaffected children. On the parental chromosomes shown below, fill in the galactosemia alleles (*A/a*) and the RFLPs (10 or 6-4) on the chromosomes to indicate linkage.

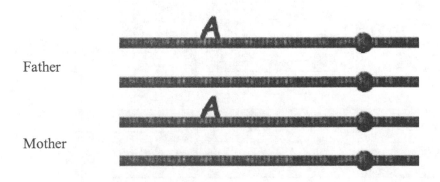

Part b

As a second example, let's look at linkage between the gene causing galactosemia and a nearby RFLP. In the pedigree above, the presence of the rare disease galactosemia is indicated by filled symbols. Using *A* and *a*, assign genotypes to each individual.

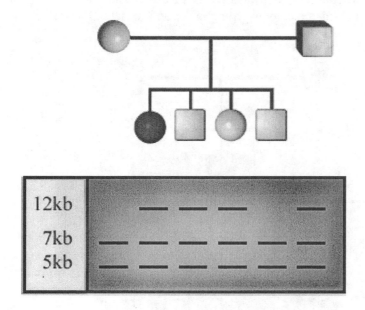

DNA was isolated from each of the individuals in the pedigree above and a probe from a linked RFLP was used in a Southern analysis to determine the genotype of each of the members of this family. On the parental chromosomes shown below indicate the linkage between the galactosemia alleles (*A/a*) and the RFLPs (12 or 7-5).

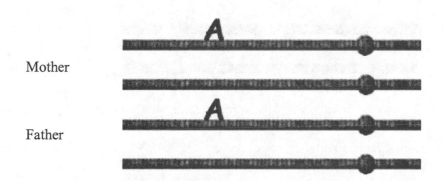

What is the probability that Child 3 is a carrier of galactosemia? (Assume the RFLP is tightly linked to the galactosemia gene and no crossing over occurs between them).

What is the probability that Child 4 is a carrier of galactosemia? (Again assume tight linkage and no crossing over.)

Problem 3

Part a

The pedigree shown below traces the inheritance of polydactyly, associated with a dominant, autosomal allele (D). Southern analysis reveals a closely linked RFLP. On the chromosomes shown below, indicate this linkage by typing in the appropriate D or d alleles and the linked 8 kb (8) or 6-2 kb (6-2) alleles.

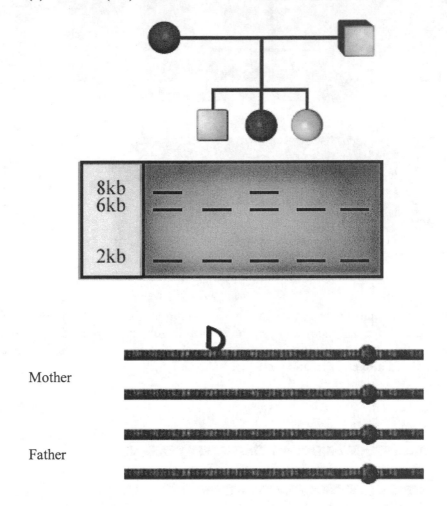

Use the Southern blot above to determine which children are heterozygous for polydactyly.

Child 1 marries a normal female and they are expecting a child. What is the probability the child will be affected?

Child 2 marries a normal male and they are expecting a child. What is the probability the child will be affected?

Would analysis of this RFLP in the fetus help them determine whether their child will have polydactyly?

Part b

Red-green colorblindness in humans is caused by an X-linked recessive allele (*c*). An STRP (small tandem repeat polymorphism) was found that is closely linked to the colorblindness gene. Assume no recombination in this problem. PCR was used to amplify this region of the genome for the individuals in the pedigree. Assign the appropriate alleles (*C*, *c* or *Y*) and linked STRPs (6 or 7) on the parental chromosomes below.

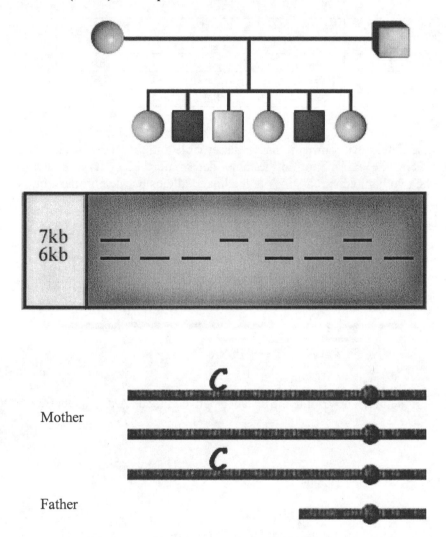

Use the gel above to determine which children are heterozygous.

Child 1 marries a normal male and they are expecting a child. What is the probability their child will be colorblind?

They find out that their baby will be a boy. Would analysis of this STRP in the baby help them determine whether their child will be colorblind?

Problem 4

You have recently identified two molecular probes (A and B) that hybridize to chromosome 12 in yeast. Although the loci are linked, you suspect they may be far enough apart to measure recombination. To test this, you mate two haploid strains to produce a diploid, which is then induced to undergo meiosis. You examine 100 meiotic haploid spores by Southern blotting and find the results shown on the next page.

Pattern	# spores	Probe A	Probe B
1	44	8 kb, 1kb	4 kb
2	7	9 kb	4 kb
3	42	9 kb	4.4 kb
4	7	8.1 kb	4.4 kb

You notice immediately that four patterns are discernable and that these are found at different frequencies. Using the information in the table, determine the genotype of the diploid cell formed from the two haploid parents. Write the linked alleles on the chromosomes below.

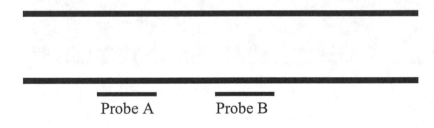

Probe A   Probe B

Using the data in the table, determine the distance between locus A and locus B.

Problem 5

Individuals with Duchenne's muscular dystrophy (DMD) are missing an important structural "glue" of skeletal muscle. This makes them susceptible to muscle tears and leads to the progressive death of muscle tissue. Many patients show pseudo-hypertrophy (false muscle enlargement), especially in the calves, as muscles die and are replaced by fat and connective tissue. Almost all are confined to a wheelchair by the age of 10, and most die in their twenties, as their respiratory muscles fail.

Although the gene leading to DMD has been cloned and sequenced (dystrophin located on the X chromosome) and many mutations can be detected directly, not all alleles have been identified. In these cases, a linked polymorphism can be used to determine probabilities. Suppose in the following family an STRP located 6 map units away has been identified.

Answer the following questions taking into account any recombination that may occur. Note that all known disease alleles are recessive, appearing primarily in males.

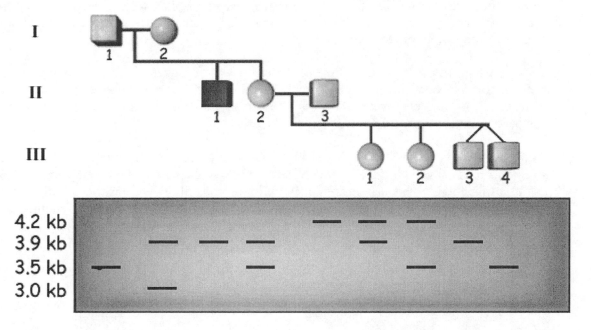

Roya (II-2) and Fardad (II-3) are in genetic counseling, after finding out she is pregnant with twin boys. Roya's brother has DMD, and it is believed to be inherited from their mother. What is the probability that Roya is a carrier? Remember, the polymorphism shown is known to be 6 map units from the DMD gene. Assume there is no crossing over in the mother I-2 leading to the child II-1.

Roya has two unaffected girls. What is the probability that her daughter Sonya (III-1) is a carrier for DMD?

What is the probability that her second daughter Kate (III-2) is a carrier for DMD?

Roya is concerned about her twin boys. The first question she asked the genetic counselor was if the twins were identical or fraternal. Based on the gel above, how would you answer this question?

What is the chance the first twin (III-3) has DMD?

What is the chance the second twin (III-4) has DMD?

Problem 6

Megan, a nineteen year old, was found severely beaten off the side of a small rural road by a patrol officer. She was taken immediately to the nearest hospital, where she was treated for her injuries and it was discovered that she had also been raped. A semen sample was recovered and stored for DNA analysis.

Two men seen in the area at the time of the crime were brought in for questioning. They claimed to have been together all through the night and knew nothing about the young woman. Blood samples were taken from both men and four STRPs were examined and compared with the semen sample.

Southern analysis of both the suspects' blood samples (1 and 2) and the semen sample (S) was performed using well characterized STRPs. Allele frequencies determined using the FBI databases for the U.S. population were then used for probability determinations (only alleles *A*, *B*, and *C* shown). It is assumed these alleles are present in Hardy-Weinberg equilibrium. From the data below, can either suspect be excluded?

STRP 1   1   2   S

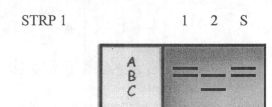

Frequency in US Population

Allele	Frequency
A	1/3
B	1/9
C	1/15

What is the probability that an individual would have the genotype consisting of STRP alleles *A* and *B*? Does this prove that Suspect 1 committed the rape?

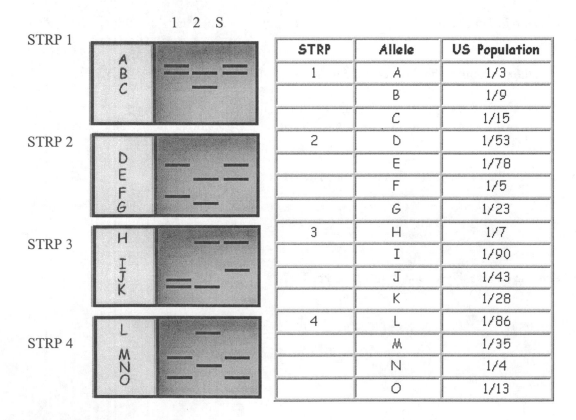

STRP	Allele	US Population
1	A	1/3
	B	1/9
	C	1/15
2	D	1/53
	E	1/78
	F	1/5
	G	1/23
3	H	1/7
	I	1/90
	J	1/43
	K	1/28
4	L	1/86
	M	1/35
	N	1/4
	O	1/13

An additional three independent STRP loci were typed (as shown above). Looking at this additional data, did Suspect 1 rape the victim?

What is the probability of an individual in the United States having the set of alleles found in the semen sample for all four STRPs?

In a population of six billion (the current population of the earth), how many individuals will have this pattern?

Medical Genetics

Goals for Medical Genetics:
1. Understand differences in classification of genetic diseases as chromosomal, single gene, or multifactorial in origin.
2. Understand how to use pedigree information to determine the mode of inheritance.
3. Interpret karyotypes to identify genetic diseases associated with chromosomal disorders.
4. Be able to understand and use in diagnosis molecular genetic techniques (ASO, FISH, and PCR) that recognize mutant DNA causing genetic diseases.

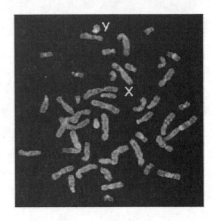

In this section, you will use case studies to see how the genetic techniques you have studied are used in medicine to help doctors form a diagnosis. To aid you in your diagnosis, the patient's clinical description and family history will be available to you. You will be able to perform various chromosomal and molecular genetic tests. At any point in your investigation, you can use the genetic reference database, and whenever you are ready, you may submit a diagnosis. You may want to take notes along the way to keep track of the information you have gathered.

Case I

Frank and Mary Smith asked that their six-year-old son George be seen by a pediatrician. His parents report that that he was slower than their other children to make his developmental milestones. They had attributed his delays to a heart defect that he had when he was born that had required surgery in infancy. He has now started school and is having trouble because he is acting out and having some behavioral problems. They are concerned that he is not ready for school.

The pediatrician is concerned about the developmental delays but notes that George's language skills appear normal or even advanced for a six year old. He notes facial dysmorphisms (unusual facial features) that seem different from his family, including full lips and puffy eyes. The pediatrician suggests that genetic tests be performed to rule out the following conditions for which further information is given:

> Angelman syndrome
> Di George syndrome
> Down syndrome
> Fragile X
> PKU
> Prader-Willi syndrome
> William syndrome

Case II

Han Chen and Yuh Nung Lee are referred to the Westside Fertility Clinic after a series of miscarriages. Han Chen is 43 and her husband is 38. Neither reports any major health problems. Han Chen was born in Taiwan. Four brothers and sisters survive in her immediate family. Two siblings died shortly after birth and one was stillborn. Yuh Nung is an only child with no family history of infant mortality.

The clinic first evaluates Yuh Nung's sperm count and finds it is normal. During this evaluation Han Chen became pregnant, but loses the fetus after three months. Fetal tissue was obtained and examined for chromosomal abnormalities in the cytogenetics laboratory. Refer to the chromosomal section under disease reference for the list of chromosomal disorders that are considered.

Case III

Sara and David Goldenstein have brought their daughter Rebecca in for her six-month checkup. They are concerned because she has stopped gaining weight, though she seems to have a healthy appetite. She has had frequent and numerous colds and suffers from diarrhea. Sara and David are first cousins, but there is no other family data that is useful.

The physician finds no evidence of mental retardation. Based on the relatedness of the parents, a recessive single Mendelian gene is suspected of causing the symptoms. Refer to the single gene section under disease reference for the list of disorders that may be considered in this case.

Case IV

Jan de Broek (age 60) was referred for medical examination by social services after his arrest for vagrancy and disorderly behavior. Social services reports that he has lost his job and apartment and is currently homeless. Jan's family was contacted and states that his behavior seems similar to both his father's and uncle's at this age (they are both deceased).

The physician records several of the symptoms associated with senile dementia. These include episodes of severe irritability, forgetfulness, anxieties, ataxia, and alcohol abuse.

Behavioral changes are among the hardest to diagnose; age-related causes may include alcoholism, Alzheimer disease, and Huntington disease. Schizophrenia is usually an early onset disorder, but should be considered if the patient history is uncertain. Further information is given in the disease reference list.

Molecular Biology

Gene Expression

Goals for Gene Expression:

1. Understand the molecular mechanisms by which the genetic information in a DNA sequence is converted to an amino acid sequence, or protein.
2. Be able to compare the similarities and differences between prokaryotic and eukaryotic transcription and the proteins involved in this process.
3. Understand the mechanism of splicing and how introns and exons are mapped in eukaryotic genes.
4. Understand how RNA is translated into a protein and how changes in a gene's sequence can give rise to defective proteins.

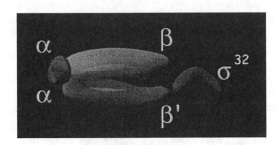

Problem 1

RNA Polymerase holoenzyme in the cell contains a sigma factor that is 70 kD (σ^{70}). However, there are several other sigma factors that recognize different promoter sequences and respond to specific cellular signals. One such sigma factor (σ^{32}) is activated in response to heat shock and recognizes promoters encoding chaperone proteins and proteases. Expression of these genes helps the cell deal with the heat denatured proteins to prevent further damage.

Comparison of promoter sequences found near genes that are transcribed under heat shock conditions revealed the recognition sequence of sigma-32. Shown below are ten such promoter sequences. Determine the best consensus sequence for this promoter.

GCCTATATA
GCCCAACTT
CCCCATGTA
CCGCATTGA
CGCCACGTA
CCCGCTATT
CGCCATCTA
ACTCTTTTT
CCCTAGATA
CGCCATGTA

Promoters that have sequences that are a good match with the consensus bind RNA polymerase more often and lead to an increased amount of transcription. These are considered strong promoters. Which promoter sequence above has the best match to the consensus sequence? Which promoter would be considered the weakest promoter?

5' AGCCTAGCTCCATATAGAACGATCATCTAAG 3'
3' TCGGATCGAGGTATATCTTGCTAGTAGATTC 5'

You have recently found a new heat shock gene above and decide to use the consensus sequence to give a first approximation of the transcription start site (+1). Which nucleotide in the sequence above represents +1?

Write the appropriate substrates in a chemical equation below to describe formation of the first phosphodiester bond from this promoter.

Which end of the mRNA has the triphosphate group?

Only one of the two DNA strands is used as a template for RNA synthesis. Which strand is used in this example?

Problem 2

Exon-intron structure of genes can be determined in a number of ways. One method involves the comparison of <u>cDNA</u> with genomic DNA. This can be done either by DNA sequencing or Southern analysis. In this problem, the gene structure of calcitonin will be examined by Southern analysis.

Calcitonin is a peptide hormone synthesized by the thyroid gland and serves to decrease circulating levels of calcium and phosphate. This is achieved primarily by inhibiting bone decalcification (or resorption).

mRNA was isolated from human thyroid cells and converted to cDNA using reverse transcriptase. Isolation of the calcitonin cDNA was done by hybridization with the calcitonin genomic DNA (previously cloned). Ten subclones from the regions of the genomic DNA shown below were isolated and labeled for use as probes (1–10).

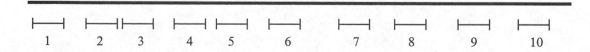

In order to determine where the calcitonin gene is located in the original genomic clone, Southern blots of the thyroid cDNA clone were hybridized with each of the ten probes (lane t, vector plus cDNA insert). A control of total genomic DNA digested with Not1 is included in lane g. See CD-ROM for the hybridization results.

The Poly-A site was found to be on the right side of the genomic DNA above, orienting the gene in a left to right direction. Use this information and the results from the Southern analysis to create an exon-intron map for the calcitonin gene.

Surprising results were observed when the same probes were used in conjunction with a cDNA clone isolated from neurons. Go to the CD-ROM to observe the results of Southern analysis on the neuronal cDNA clone (n) versus genomic DNA (g). Additional questions are available on the CD-ROM.

Problem 3

Hemoglobin is a tetrameric protein ($\alpha_2\beta_2$) that functions to carry oxygen in all vertebrates and some invertebrates. Each subunit of hemoglobin is associated with a prosthetic group, iron containing heme (in white above), which provides the ability to bind oxygen. Disorders associated with alterations to, or the disrupted synthesis of, hemoglobin are the most common genetic diseases in the world.

The β-globin gene is a relatively small gene, composed of three exons and two introns. One major class of hereditary disorders associated with hemoglobin is characterized by alterations in the amino acid sequence of β-globin but not the amount (structural variants). A second major class is characterized by a reduced level of hemoglobin (thalassemias), due to either mutations affecting transcription or splicing. In the following problem, β-globins were obtained from mutant hemoglobins. Use your knowledge of gene expression to help determine the cause of each of the mutations described.

Blood samples were collected from individuals homozygous for several distinct hemoglobin disorders. Protein from each sample was electrophoresed through either a native polyacrylamide gel or an SDS gel (shown below) and the separated proteins transferred to nitrocellulose. β-hemoglobin was then detected using antibodies (Western analysis).

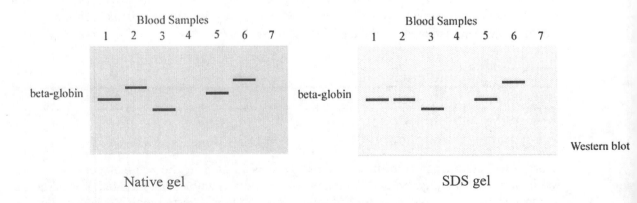

Native gel   SDS gel

The first lane (blood sample 1) contains wild-type β-globin. Which lanes from 2–7 above represent structural changes in β-globin?

Compare the sample in lane 2 on a native versus SDS gel. Do these results suggest a change in the length of the polypeptide or the charge of the polypeptide?

Compare the sample in lane 3 on a native versus SDS gel. Do these results suggest a change in the length of the polypeptide or the charge of the polypeptide?

DNA and RNA samples were then examined by <u>Southern</u> and <u>Northern blotting</u> to detect changes in the DNA or RNA.

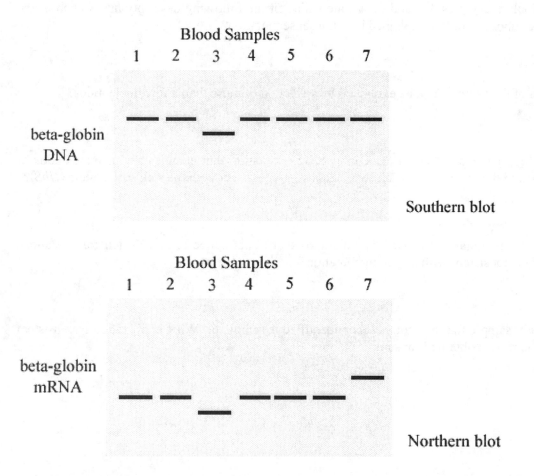

Again, the first lane contains a wild-type sample. To begin, let us examine sample 2 in detail. You have already determined the β-globin polypeptide differs from the wild type by a charge change. Does the Southern blot detect any difference in sample 2 from wild type?

Does the Northern blot detect any changes in mRNA length in sample 2?

DNA from sample 2 was analyzed by dideoxy sequencing and compared to the wild-type gene. A single mutation was found in exon 1, as shown below (only the nontemplate strand is shown). Use the genetic code to determine whether this mutation represents a nonsense mutation (stop codon), missense mutation (amino acid substitution) or a frameshift mutation (insertion or deletion).

Wild type   ATG GTG CAC CTG ACT CCT G̲AG GAG AAG TCT GCC

Sample 2   ATG GTG CAC CTG ACT CCT A̲AG GAG AAG TCT GCC

Now look at samples 3–7 and determine which fit the following descriptions. Which of the samples above could be explained by a nonsense mutation?

Which of the samples can be explained by a deletion of more than a few nucleotides?

<u>Sickle-cell anemia</u> is known to be caused by a mutation that changes a glutamate codon at amino acid 6 to a valine. Which sample above is consistent with the sickle cell allele (*HbS*)?

Many thalassemias are caused by mutations that effect splice sites. Which sample above would be consistent with a splicing mutation?

The only sample that has not yet been identified is sample 6. What explanation is consistent with the results obtained for sample 6 above?

Molecular Biology

Gene Cloning

Goals for Gene Cloning:

1. Review the use of restriction enzymes and be able to use restriction enzymes to make a physical map of a clone.
2. Review basic cloning techniques, including vectors, library construction, and screening techniques.
3. Review the DNA sequencing, Southern analysis, and PCR amplification.
4. Be able to design a cloning strategy, beginning with understanding what is to be cloned, why it is to be cloned, and what will be done with the clone. The cloning strategy consists of the origin of DNA (genomic or cDNA library), choice of cloning vector, and choice of screening technique to recognize clone.

(-) electrode

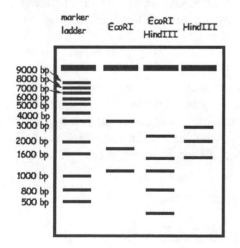

(+) electrode

Problem 1

A restriction map provides information as to the presence and position of one or more restriction sites. In many ways it is like the road map to a DNA molecule, and it is used in many recombinant DNA techniques.

Digestion with a single enzyme can provide information about the number of restriction sites and the distance between them, but the order of the sites cannot be discerned. However, if a second enzyme is used both alone and in combination with the first enzyme, a map can be constructed.

In this problem, you will determine the restriction map for a linear DNA sample using the enzymes Eco RI and Hind III. Three samples of DNA are first digested with either Eco RI, Hind III, or both Eco RI and Hind III. The resulting fragments (restriction fragments) are separated by size using gel electrophoresis.

The results are shown here.

Since the DNA fragments in each lane are derived from the same linear DNA fragment, the sum of the size of fragments in each lane should be the same and reflect the size of the original DNA fragment. How many base pairs are in the original DNA molecule?

Focus first on the Eco RI single digest. Since we know our sample DNA is linear, how many EcoRI sites are present in the DNA?

When the same DNA sample is digested with Eco RI and Hind III, five fragments are generated. Let's look at this one band at a time. Since the 3000 bp Eco RI fragment is not present in the double digest, it must contain at least one Hind III site. Which fragments in the double digest add up to 3000? Which bands in the double digest add up to 1800? Now there is only one fragment left in the double digest. This is the same size as one of the Eco RI fragments in the single digest. It must correspond to the band in the Eco RI lane.

Now, let's focus on the Hind III digest. The largest band, 2500 bp, disappears in the double digest. Which two bands in the double digest add up to 2500 bp? Which two bands in the double digest add up to 2000 bp? That leaves just the 1500 bp Hind III fragment, which is also present in the double digest. This means that there are no Eco RI sites inside this fragment and may indicate the end fragment on a linear DNA.

Since this is a linear DNA fragment, let's start with one of the fragments that is the same in both the single and double digest. This could be an end fragment. Although we could start with either the 1200 or 1500 bp fragment, let's begin with placing the 1200 bp fragment on the left side and build the map from there. Create a map using the information gained from the single and double digests above.

Problem 2

You have just identified a patient who appears to have an altered hemoglobin protein (β-subunit). Analysis of the patient's blood by Southern, Northern, and Western blotting reveals no detectable change of the DNA, a larger RNA, and smaller protein. You surmise that the mutation affects hemoglobin splicing. With a clone of the wild-type hemoglobin in hand, and cultured skin cells from the patient (as a source of DNA or RNA), select a type of insert, a type of vector, and a method in which to identify and study the mutant hemoglobin gene.

Problem 3

After many months in the laboratory, you have finally obtained a pure preparation of DNA polymerase from a novel fungus. However, you have a very limited amount and cannot do all of the biochemical experiments you would like. Your advisor suggests that you use your purified polymerase to clone the gene encoding the polymerase. After consulting several other graduate students, you come up with a plan. Select a type of insert, a type of vector, and a method to identify the DNA polymerase gene.

Problem 4

Polydactyly is known to be caused by a dominant allele and is characterized by extra fingers and/or toes. Recently, an RFLP has been identified which is linked to polydactyly and you decide to use this information to clone the gene. Determine what type of insert you would use in making your library, what type of cloning vector would be optimal, and how you would go about identifying the correct clone.